Dispelling Myths about Water Services

Dispelling Myths about Water Services

Edited by

Tapio S. Katko, Jarmo J. Hukka,
Petri S. Juuti, Riikka P. Juuti and
Eric J. Nealer

Illustrations: Pertti O. Väyrynen

Published by   **IWA Publishing**
          Unit 104–105, Export Building
          1 Clove Crescent
          London E14 2BA, UK
          Telephone: +44 (0)20 7654 5500
          Fax: +44 (0)20 7654 5555
          Email: publications@iwap.co.uk
          Web: www.iwaponline.com

First published 2025
© 2025 IWA Publishing

Disclaimer
The information provided and the opinions given in this publication are not necessarily those of IWA and should not be acted upon without independent consideration and professional advice. IWA and the Editors and Authors will not accept responsibility for any loss or damage suffered by any person acting or refraining from acting upon any material contained in this publication.

British Library Cataloguing in Publication Data
A CIP catalogue record for this book is available from the British Library

ISBN: 9781789064155 (paperback)
ISBN: 9781789064162 (eBook)
ISBN: 9781789064179 (ePub)

Doi: 10.2166/9781789064162

This eBook was made Open Access in September 2025.

Contents

Chapter 4
Rules of water services: how to promote teamwork? **87**

Chapter 5
The future of water services **141**

doi: 10.2166/9781789064162_ix

Foreword

In this new world of instant information, myths and disinformation shape our behaviour and make it appear that the rational world of the Enlightenment is moving away from us. Here, the authors resist this trend by undertaking a comprehensive explanation of the myths surrounding the existential subject of water services. They are well qualified to undertake this difficult task amidst a wealth of available information and disinformation. Four of them are colleagues at Tampere University in Finland and one is from the University of South Africa in Pretoria. All have extensive experience in water services across a range of systems in different countries, relating to all income categories and various stages of development.

Their task is challenging because people use water in many different social ways and water management also involves business and nature. Water is seen as a connector as it is essential for so many needs in society. The most visible aspects of water management on a global basis have been assigned to Sustainable Development Goal six, which is 'To ensure access to water and sanitation for all' [https://www.un.org/sustainabledevelopment/water-and-sanitation]. These two elements are the main areas of focus by the authors in their coverage of community water supply and wastewater systems.

Sanitation involves the additional aspect of providing toilet and washing facilities for people within their living spaces; while this is a major global issue, it is somewhat outside the remit of water resources management. The authors could have included stormwater, but it was considered better to maintain a sharp focus on water and wastewater.

The core of the book comprises a discussion of 21 myths, organised into four chapters. These are illustrated with attractive figures and drawings, many that are unique to this work. They are also supported by extensive references. After the introduction in chapter one, chapter two considers the origins of water and where it is used. The third chapter is about water services and this is followed by a discussion of rules or institutional arrangements in chapter four. The chapter five explores the myths themselves and looks to the future. This is not an unexpected direction to take as the authors have an interest in studies relating to futures, which contend that while we cannot predict the future, we can consider it.

The myths chosen are important for an understanding of the overlapping spheres of water services and water resources management. The distinction between the two concepts is addressed in the discussion of the first myth, i.e. that 'Water resources and water services mean the same thing'. The authors explain that water resources management, comprising use, planning, and governance, can involve situations ranging from international treaties and transboundary waters to management of water bodies, whereas water services are governed at the local level where people rely on them. Personally, I have encountered this myth many times over the years and support the distinction made here. My own explanation focuses on water resources management as an overall approach that addresses all uses and occurrences of water, whereas water services are a subset of public services, which are the essential support systems provided by governments.

A further frequently occurring myth is 'Water is a basic human right, and therefore, it should be free'. In discussions with my students, we find that reaching an understanding of these issues requires us to dive into the complex concepts of human rights, equity, affordability, and the needs of management to ensure these complicated water systems operate reliably. One explanation of this that I often offer students is that water should be free, but who pays to keep the systems running? From here, one must dive into the world of managing water infrastructure and organisations, which is more complex than generally realised.

Taken together, the 21 myths contained in this book offer a useful and comprehensive view of current issues in the water services sector. Thinking about these, one cannot help but be impressed by the complexity of the 'rest of the story', a line used by the famous US radio host Paul Harvey many years ago to explain societal issues. This book helps us to focus on the rest of the story of water services and is a small step toward achieving the goals of Sustainable Development Goal 6.

Professor Neil Grigg
Department of Civil and Environmental Engineering
Colorado State University, Fort Collins

Preface

Water occurs in many forms in our daily lives and due to its diversity, can mean a range of different things to different people. Some of these meanings can develop into myths rather than being based on facts.

We all use water several times a day. Water, sanitation, and health are often taken for granted in a well-functioning society and built environment. However globally, close to one thousand children die each day due to unsafe water and lack of adequate hygiene. Historically, in what are now today's high-income countries, water and sanitation conditions were very challenging until a change occurred around the early twentieth century. The inception of water supply and wastewater systems gradually improved living conditions, health, security, and the overall standard of living.

In this book, the concept of water services covers both community water supply and wastewater systems or sanitation. Although this could include stormwater management, this is not covered comprehensively and is mentioned only occasionally.

Through this book, the authors wish to dispel common myths related to water services, while simultaneously raising questions of concern relating to water services. People do not necessarily think about what kind of issues and challenges are related to water services and systems. The original book was published in Finnish; this English version has been completely rewritten and includes examples and references from various countries and continents plus a few cases from Finland.

The COVID-19 pandemic in particular reminded us of the necessity and indispensable value of water services. For example, we learned that viruses can be reduced by washing hands with warm water and soap. This requires operational water services, which are not yet available for all. Sadly, we have also witnessed the value of water to nations currently at war.

Regarding the Sustainable Development Goals (SDGs), number six (SDG6) on clean water and sanitation has many connections with other goals. In practice, none of the 17 goals can be achieved without reaching SDG6. Adapting to and mitigating climate change also requires more sustainable governance and management of water resources and services.

Several colleagues have contributed to the preparation of this book with their advice and comments on selected parts or the whole manuscript: Andreas Angelakis, Ari Ahonen, Eero Arola, Ritva Britschgi, Neil Grigg, Laura Inha, Todd Jarvis, Ari Kangas, Seppo Knuuttila, Japhet Koroos, Toivo Lapinlampi, Vuokko Laukka, Richard Little, Alexandros Makarigakis, Harri Mattila, Olli Niemi, Reijo Oravainen, Kenan Ozekin, Mikko Perkiö, Kenneth Persson, Pekka Pietilä, Niko Putkinen, Eija Raimovaara, Jari Rintala, Osmo Seppälä, Arto Suominen, Mahendra Umare and Eija Vinnari. In addition, we offer our sincere thanks for the open-ended reflective essays written by six experts from four continents contained in chapter six.

The advice and assistance from Katharine Allenby, Mark Hammond, Andrew Peart and Joe Pilbrow from IWA Publishing, London, are gratefully acknowledged. The language of the manuscript was checked by Professor Eric Nealer from the University of South Africa, Pretoria, with final polishing by Dr Julie Fisher from High Peaks Editing, UK.

The writing and dissemination of this book has been supported by grants from the Association of Finnish Nonfiction Writers, the Waldemar von Frenckell Foundation, the Middle Course Foundation, the Nessling Foundation, the Research, Development and Innovation Cluster (VEPATUKI) and Tampere University (TUNI) library. Our heartfelt thanks are given for this support.

This book is dedicated to the far too many billions of people around the world who still lack sustainable water services.

Tampere, Finland
World Water Day, 22nd March 2025
The authors

Endorsements

'An invaluable analysis of an extremely important but often a seriously neglected issue.'
Risto Isomäki
Creative and professional writer, Finland

'Water has been around us throughout the entire history of humanity and through all our life. It is part of all the cultures around the world, each country and community has their own views and ways to manage the provision of water services. Naturally, myths have emerged about how these services should be managed. But what are those myths and, most of all, why are they myths? You will need to read the book (and sometimes allow yourself to be surprised) to find the answers, and inspire you to reflect on what other myths might exist.'
Blanca Elena Jiménez Cisneros
Ambassador of Mexico in France and Monaco, Senior Researcher at the Autonomous University of Mexico and member of the International Water Association Governing Board

'There can hardly be more important social innovations than those that address water supply or sanitation. Their impact on global health, resilience in both urban and rural areas, and sustainability generally cannot be overestimated. The book is so timely and important, both for those directly involved in resource governance and for those who want to understand how water services will shape our futures and the world.'
Professor Alf Rehn
University of Southern Denmark

'The authors discuss several well-informed avenues of thinking to demythologise and promote effective water services governance and management for the futures. It is an exceptional read for anyone interested in local and international water governance and management.'
Professor Johann Tempelhoff
North-West University, South Africa

doi: 10.2166/9781789064162_0001

Chapter 1

The importance of water services

'*We must learn to live together as brothers or perish together as fools.*'
(Martin Luther King, Jr.)

'*The best way to find yourself is to lose yourself in the service of others.*'
(Mahatma Gandhi)

1.1 INTRODUCTION

Since ancient times, the need for fresh, safe and healthy water has resulted in the development of various kinds of water supply and sanitation systems. From early history, civilisations have developed water purification devices and treatment methods. The need for fresh water has influenced individual lives as well as whole societies.

In any society, water and wastewater systems are of fundamental importance to the development of communities and the wellbeing of both people and the ecosystem. Unfortunately, this fact has been reinforced by the COVID-19 pandemic, by all manner of natural disasters, and by recent armed conflicts around the world. In such situations, clean water and sanitation are among the first things that need to be organised.

Mornings are a great time to reflect on the often invisible individuals who work behind the scenes, powering our societal institutions and networks of services. These dedicated professionals – including plumbers, civil servants, nurses, teachers, lawyers, police officers, firefighters, electricians and many others – are the unsung heroes keeping our communities running smoothly. However, the World Economic Forum and World Resources Institute (2023) estimated that each year, 25 countries or one-quarter of the global population face extremely high levels of water stress, and at least half the world's population live under these conditions for at least one month a year.

According to the United Nations World Water Development Report 2018 (UN Water, 2018a), nearly six billion people could potentially suffer from water scarcity for at least one month a year by 2050. Boretti and Rosa (2019) suggested that this number may be an underestimation due to increasing water demand, reduction of water resources, and increasing water pollution, all of which are driven by dramatic population and economic growth. Improvements in science and technology may alleviate these problems, yet, they argued that effective public policy is much more urgent, together with proper regulation and enforcement.

Water services (community water supply and wastewater services) for all is one of the greatest challenges of the twenty-first century, especially in low-income countries, where 66% of the population is projected to live in cities by 2050 (UNDESA, 2014, p. 1). While high-income countries are facing the difficulties of maintaining and rehabilitating their existing systems, in low- and middle-income economies, the question is more about the lack of access to and the unviability of water and sanitation services.

In many high-income societies, wastewater treatment plants are often among the greatest environmental investments for communities. Unfortunately in the global context, as much as 80% of the world's wastewaters still remain untreated (UN Water, 2018b), and consequently, 1.8 billion people use a drinking water source contaminated with faeces. A more recent study by Jones *et al.* (2021) suggested that 48% of global wastewaters are released untreated into the environment. Other estimates show that only 10% of the world's wastewaters are treated efficiently. According to UNU–INWEH (n.d.), 52% of wastewaters are treated globally, while rates vary drastically between high-income (74%), upper-middle income (43%), lower-middle income (26%) and low-income (4.3%) countries. In any case, the global need for far more treatment capacity is acute.

A paradox can be seen if we look at the worldwide context. In communities with well-functioning and managed systems and services, people tend to take these for granted and fail to value safe drinking water and proper sanitation as they should. Too often it is argued that water and wastewater systems are invisible since a major part of their infrastructure lies underground. For most people, use of this infrastructure begins at the start of the day by washing faces, hands and teeth and defecating (emptying our bowels using the toilet). In comparison, if one does not have access to these services, this is felt through the substantial efforts needed to carry water or buy it from vendors for a high unit price, the need to use shared or unimproved sanitation facilities, or even practise open defecation. Indeed, fees paid for water are relatively highest among peri-urban dwellers in low-income countries although their voices are not heard.

One of the key questions is how to consider water and water services as a commodity or resource. Water services in any community constitute a natural monopoly, a concept first introduced by John Stuart Mill in 1848 (Sharkey, 1982). Therefore, it is not feasible to construct more than one network in a single area. In many countries, the first water and wastewater works were implemented through private concessions until 1900, with most of them later transferred into municipal ownership. This continued until 1989 when the UK Government decided to privatise regional public water authorities in England and Wales.

Privatisation of water services was, however, soon followed by a global movement of remunicipalisation. From 2000 to 2015, a total of 235 urban water services in 37 countries were taken back into public ownership (Kishimoto & Petitjean, 2017). By September 2024, this number increased to 377 cases globally (Public Futures, 2024). It is currently estimated that approximately 90–95% of customers worldwide receive their water from publicly owned or operated utilities. The question of privatisation of water services is discussed further under Myth 17 in chapter four.

It is possible to abandon a long-term perspective during crises such as the COVID-19 pandemic; the resulting biased perspective can make us see only the world's inevitable destruction. Philosopher and academic, Georg Henrik von Wright (1916–2003), wrote several essays in the 1970s published in the book 'Humanism as life stance' (von Wright, 1981). He pointed out how each generation sees the particular problems of their time to be the greatest, however, he reminded us of the necessity to reassess the possibilities available to us to be able to regain control and to direct technological development wisely (von Wright, 1981, p. 191). The historian Yuval Noah Hararí's view is that people tend to overemphasise the importance of their own era including any gains and losses (Nieminen, 2020).

In high-income societies, it is easy to wash hands with soap and warm water. In low- and middle-income economies, it is the reverse; in 2022, two billion people still lacked basic hygiene services, including 1.3 billion with limited services and 653 million with no facility at all (UNICEF & WHO, 2023).

The UN World Water Development Report 2021 pointed out the value of water. The failure to fully value this in all its different uses is considered to be a root cause or symptom of the political neglect of water and its mismanagement. All too often, the value of water is not prominent in public decision making (UN Water, 2021).

The 2030 Agenda for Sustainable Development was adopted by all United Nations Member States in 2015 (United Nations, 2015). This provides a shared blueprint for peace and prosperity for people and the planet, now and into the future. At its core are the 17 SDGs, which call all countries –developed and developing – into a global partnership. SDG6 has specific focus on clean water and sanitation although almost all the other goals are also connected to water (Figure 1.1).

According to Grigg (2016), water is a sector but also a connector of many other sectors and, therefore, it is particularly important and even unique in terms of sustainable development. Ending poverty and other deprivations must go hand in hand with strategies that improve health and education, reduce inequality, and spur economic growth – all while tackling climate change and working to preserve our oceans and forests (UNDESA, n.d.). Accurate data on global trends in urbanisation and city expansion is critical for assessing current and future needs relating to urban growth and for creating public policy priorities to promote inclusive and equitable urban and rural development. (UNDESA, 2014)

Water is largely connected to the question of overall wellbeing, which according to Maté (2022, p. 3) 'must move from the individual to the global in

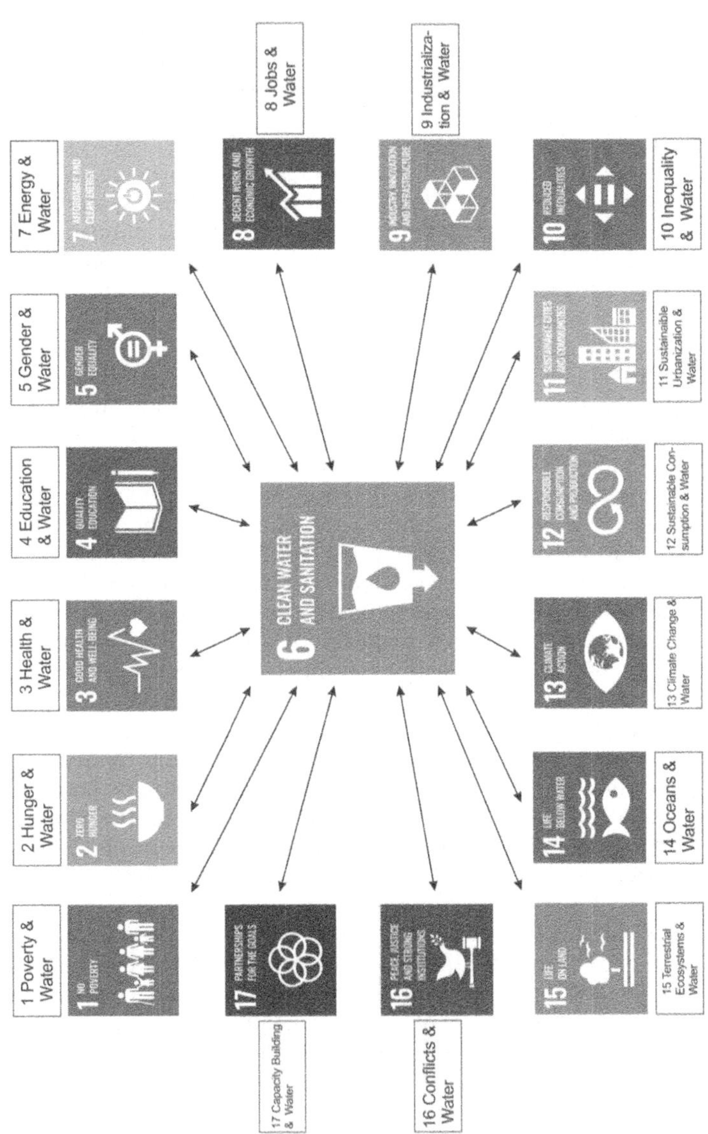

Figure 1.1 The 17 sustainable development goals of the United Nations: SDG6 on clean water and sanitation and its connections to other goals. *Source:* UN (n.d.), illustration: T. Katko 2024.

every sense of that world'. Berman (2001, pp. 54–65, cited by Maté 2022, p. 3) refers to global capitalism that has become a 'total commercial environment that circumstances an entire mental world'.

1.2 WHAT IS THIS BOOK ABOUT?

This book deals with questions that people regard as problematic in terms of community water and wastewater services, referred to here as 'water services'. It is based on the findings and experiences of public speaking and delivering lectures, especially by the first author, over a number of years to a variety of audiences, mainly in Finland but also elsewhere.

Additionally, the book incorporates research findings and cites many international references. It is meant for a wide audience including public decision makers, citizens, schoolchildren (senior), students, the media and sector professionals. Our intention has been to use language that avoids professional jargon as much as possible. Besides, the authors wish to promote critical thinking – to find a balance between academic and popularised text while at the same time promoting dialogue on water services and their importance.

The book covers a total of 21 selected myths. These are based on qualitative empirical evidence (Costa, 2025) from discussions with various audiences. In 2010, it was noted that the questions raised by different audiences started to recur. Most of these myths have been raised by members of the general public but also by sector professionals, civil servants and decision makers. Although work to gather these myths started before then, they were collected more systematically between 2016 and 2020 through various presentations to the general public and student groups, particularly in the field of civil and environmental engineering and health sciences. The authors have used their field experiences and research findings from several countries, for example, in Africa, Asia, Europe and North America, to explore these myths or what the authors see as 'reality'.

The authors' first book published in Finnish in 2022 on 'Myths on Water Services' focused on the situation in Finland and was well received. After this encouragement, the present version was completely rewritten in English, the authorship was expanded, mainly international references from 43 countries were used, and six reflective essays from various parts of the world were received. Consequently, it is believed that the book now has a large potential global audience among English readers. It is a deliberate open-source publication, freely downloadable, while available also in print. It is our hope that it will be widely distributed and used.

The research material is predominantly based on more than 120 research projects conducted through the inter- and multidisciplinary Capacity Development in Water and Environmental Services (CADWES) research team, established in Tampere in 1998 by the Finnish authors who were located at two local universities – Tampere University and Tampere University of Technology. It defined its mission 'to produce usable knowledge and education, based on multi- and interdisciplinary research, on the evolution and development of sustainable use of water services and water resources in a wider institutional context: organisations, governance, management, economics, legislation, policy, rules and practices' (CADWES, a). This water services mission involves a process of putting pieces together as described by Maté (2022, p. 9). It aims to give a broad understanding of systems and their management and the related institutions and governance.

The authors have also utilised the findings from a variety of educational activities, including the continuing education programme 'Water Services Leadership and Development (WASLED)' which has taken place seven times between 2009 until 2024 (with 123 participants to date) and the annual conferences of the UNESCO Chair in Sustainable Water Services since 2012 (CADWES, b).

Through the presentation, exploration and discussion of these myths, the authors present answers to the following key questions:

(i)   What kind of myths prevail among the public regarding water and sanitation services and why?

(ii)   How do myths impact water services regulatory governance, provision and production?

(iii)   How could better understanding of myths be achieved in order to clarify and dispel them?

The major part of water services infrastructure is physically invisible because it is located underground. Yet the invisibility of water services is a myth in itself, since everybody uses these services several times a day, and sometimes even at night. Once they become non-operational, their real value is soon recognised. If pipes were to burst, people would experience inconvenience and even headaches due to the disruption of their daily routines. This leads to the question: how can the importance of this invaluable water infrastructure be more known and recognised for societal decision making? (Bennich *et al.*, 2023; Prevos, 2018). A recent survey carried out in six countries shows that people do appreciate the value of water services (Kemira, 2020).

The book is organised into seven chapters. Following the introduction in chapter one, chapters two and three focus on technological issues. The second chapter covers water resources and water use in communities (Myths 1 to 4) and the third chapter, water services as community circulatory systems (Myths 5 to 12). The fourth chapter deals with institutions and rules relating to water services (Myths 13 to 17), while chapter five looks at futures (Myths 18 to 21). At the end of each of chapters two to five are questions for further discussion. Chapter six covers requested reflections by guest experts from six countries and regions in different parts of the world, and chapter seven is the concluding remarks. Each chapter has its own list of references.

1.3 THE NATURE OF MYTHS

First, the nature of myths is discussed in a wide context, after which the specific myths related to water services are explained in the subsequent four chapters. It is probably fair to say that in any nation its historical myths are open to debate and it is likely that each generation has to create its own interpretations of these developments. As Jussila (2007) noted, new myths are born all the time.

Myths in world history are explored, for example by Sonnabend (2014). Assumptions about the Dark Middle Ages have existed for a long time and more recently, researchers have corrected these views (Anon, 2023; Gabriele & Perry, 2021). A collection of articles titled 'Galileo Goes to Jail and Other Myths about Science and Religion' elaborates on and invalidates 25 myths. Most of them originate from the late 1800s, from two single sources that are based on no scientific evidence (Numbers, 2009).

The Merriam-Webster Dictionary (n.d.) defines the term 'myth' as a popular belief or tradition that has grown up around something or someone, or a person or thing having only an imaginary or unverifiable existence.

The importance of myth versus reality has increased during the so-called post-truth era that began in the early 2000s. The Oxford Learner's Dictionaries (n.d.) defines post-truth as an adjective that is related to circumstances in which people respond more to feelings and beliefs than to facts. Post-truth or post-factual politics is a political culture in which facts are considered irrelevant. It is regarded as being influenced by the arrival of new communication and media technologies. In this era of post-truth politics, it is argued that it is easy to cherry-pick data and reach any conclusion you choose (European Center for Populism Studies, n.d.).

Here, the authors define the concept of myth as follows: *a long-lasting belief that is at least questionable from the point of research-based evidence.* The selected myths have appeared through conducting long-term research on water services and through several presentations and discussions with the general public, students and professional colleagues. In this book, the extent to which these myths are true in terms of research-based findings is explored.

After describing each myth, the authors present their views on the associated 'reality'. However, it is not proposed that these descriptions are the only possible and correct interpretations. Rather, in their understanding they are – based on available knowledge – 'interesting and possibly true', a position noted by futures researcher Pantzar (1993) relating to scientific research. Pantzar also contended that trust in futures is important especially in crisis conditions (Pantzar, 2021).

Many of the myths are 'wicked' in nature (Freeman, 2000); as complicated problems, they do not have simple and completely unambiguous answers. According to Beutler (2016), wicked problems are not solved – they can only be mitigated through an approach that emphasises empathy, abductive reasoning and rapid prototyping.

Scientific research is multidisciplinary and incoherent. The bias in favour of a positivistic, natural science approach in water research prevents satisfactory answers being found to wider water governance challenges as well as institutional and management issues. Water research should, therefore, be expanded to include diverse pluri-, cross-, inter- and multidisciplinary approaches as joint efforts, while individuals should be encouraged to seek transdisciplinarity (Hukka *et al.*, 2007). Accordingly, Karin Gardes stated that many people still primarily associate innovation with technology. She, however, argued that innovation is a much broader concept that also needs to cover governance, finance, values and culture (World Water Week, 2023). In chapter seven, the authors assess to what extent the myths presented are true or not, returning to the research questions and a discussion of the wider implications, followed by concluding remarks.

Water supply and wastewater/sanitation services are addressed in both urban and rural contexts. Stormwater management is mentioned only occasionally although in many countries it is considered to be a key part of integrated urban water management (IUWM), which is a view shared by the authors. The focus of the book is on water services produced by utilities although the importance of self-supply systems is also noted.

REFERENCES

Anon (2023). 20 Myths about the Middle Ages. https://www.medievalists.net/2023/09/20-myths-middle-ages/ (accessed 15 December 2024).

Bennich A., Engwall M. and Nilsson D. (2023). Operating in the shadowland: Why water utilities fail to manage decaying infrastructure. *Utilities Policy*, **82**, 101557. https://doi.org/10.1016/j.jup.2023.101557

Berman M. (2001). The twilight of American culture. W.W. Norton, New York.

Beutler L. (2016). What to Do about Wicked Water Problems. https://wrrc.arizona.edu/wicked-water-problems (accessed 22 March 2025).

Boretti A. and Rosa L. (2019). Reassessing the projections of the World Water Development Report. Perspective. *Npj Clean Water*, **2**(15). https://www.nature.com/articles/s41545-019-0039-9.

CADWES a. Science and Education for Water Services http://www.cadwes.com/ (accessed 22 March 2025).

CADWES b. UNESCO Chair in Sustainable Water Services. http://www.cadwes.com/unesco-chair/ (accessed 22 March 2025).

Costa D. (2025). Empirical evidence. Encyclopædia Britannica, 25 Jan 2025. https://www.britannica.com/topic/empirical-evidence (accessed 22 March 2025).

European Center for Populism Studies (n.d.). Dictionary of Populism. Post-Truth Politics. https://www.populismstudies.org/Vocabulary/post-truth-politics/ (accessed 22 March 2025).

Freeman D. M. (2000). Wicked water problems: sociology and local water organizations in addressing water resources policy. *JAWRA Journal of the American Water Resources Association*, **36**(3), 483–491. https://doi.org/10.1111/j.1752-1688.2000.tb04280.x

Gabriele M. and Perry D. M. (2021). The Bright Ages: A New History of Medieval Europe. Harper, New York.

Grigg N. S. (2016). Water as a connector among societal needs. In: Integrated water resource management. An Interdisciplinary Approach, N. S. Grigg (ed.) Palgrave MacMillan, London and New York, pp. 1–18.

Hukka J. J., Katko T. S., Mattila H. E., Pietilä P. E., Sandelin S. K. and Seppälä O. T. (2007). Inadequacy of positivistic research to explain complexity of water management. *International Journal of Water*, **3**(4), 425–444.

Jones E. R., van Vliet M. T. H., Qadir M. and Bierken M. F. P. (2021). Country-level and gridded estimates of wastewater production, collection, treatment and reuse. *Earth System Science Data*, **13**, 237–254. https://doi.org/10.5194/essd-13-237-2021.

Jussila O. (2007). Suomen Historian Suuret Myytit (Myths in Finnish History). WSOY, Helsinki, Finland, Vol. 305, p. 12.

Kemira (2020). Highlights from an international consumer survey on water. https://www.kemira.com/app/uploads/2020/10/Kemira_water_datasummary_US_FINAL-5f9bf7b272098.pdf (accessed 22 March 2025).

Kishimoto S. and Petitjean O. (eds) (2017). Reclaiming Public Services: How cities and citizens are turning back privatisation, p. 13. www.tni.org/reclaiming-public-services (accessed 22 March 2025).

Maté G. (with D. Maté) (2022). The Myth of Normal. Illness, Health and Healing in a Toxic Culture. Vermilion, London.

Merriam-Webster dictionary (n.d.). https://www.merriam-webster.com/dictionary/myth (accessed 22 March 2025).

Nieminen T. (2020) Haloo, onko siellä Yuval? (Hello, is it Yaval?) *Helsingin Sanomat* 11 April 2020.

Numbers R. L. (ed.) (2009). Galileo Goes to Jail and Other Myths about Science and Religion. Harvard University Press, Cambridge, MA, USA, pp. 1–7.

Oxford Learner's Dictionaries (n.d.). https://www.oxfordlearnersdictionaries.com/definition/english/post-truth (accessed 22 March 2025).

Pantzar M. (1993). Evoluutioteoria tulevaisuuden tutkimuksen metodina (Evolutionary theory as a futures research method). Teoksessa Matti Vapaavuori (toim.) Miten tutkimme tulevaisuutta? (How to explore futures?) Kommunikatiivinen tulevaisuudentutkimus Suomessa. Acta Futura Fennica No. 5. Tulevaisuuden tutkimuksen seura. Painatuskeskus, Helsinki, Vol. **75–88**, p. 86.

Pantzar M. (2021). 'Toivossa on hyvä elää' – Tulevaisuususko hyvän kansanterveyden ja toimivan kansantalouden edellytyksenä (It is good to live in hope – Futures belief as a requirement for good public health and operational national economy). *Futura*, **40**(2), 9–17.

Prevos P. (2018). Customer Experience Management for Water Utilities: Marketing Urban Water Supply. IWA Publishing, London, p. 2.

Public Futures (2024). Public Futures collects cases where privatisation ends and where public services, production and infrastructure are (re)created. https://publicfutures.org/about (accessed 22 March 2025).

Sharkey W. W. (1982). The Theory of Natural Monopoly. Cambridge University Press, Cambridge, UK, p. 14.

Sonnabend H. (2014). Ist Das Wirlich Wahr? (Finnish translation 2016: Menneisyyden myytit: totta vai tarua?). Reader's Digest Germany, Switzerland, Austria.

UN. Sustainable Development Goals. Communications materials. https://www.un.org/sustainabledevelopment/news/communications-material/ (accessed 22 March 2025).

UNDESA (2014). World Urbanization Prospects: The 2014 Revision, Highlights (ST/ESA/SER.A/352), p. 28. https://www.un.org/en/development/desa/publications/2014-revision-world-urbanization-prospects.html (accessed 22 March 2025).

UNDESA (n.d.). The 17 goals. https://sdgs.un.org/goals (accessed 22 March 2025).

UNICEF and WHO (2023). Progress on household drinking water, sanitation and hygiene 2000–2022: special focus on gender. UNICEF and WHO, New York, p. viii. https://data.unicef.org/resources/jmp-report-2023/ (accessed 22 March 2025).

United Nations (2015). Transforming our world: the 2030 Agenda for Sustainable Development. Resolution adopted by the General Assembly on 25 Sept 2015. A/RES/70/1.

UNU–INWEH (n.d.). Global Wastewater Status. https://unu.edu/inweh/tools-and-resources/global-wastewater-status. (accessed 22 March 2025).

UN Water (2018a). World Water Development Report 2018. Nature-based Solutions for Water, p. 3. https://www.unwater.org/publications/world-water-development-report-2018. (accessed 22 March 2025).

UN Water (2018b). Water quality and wastewater, p. 87. https://www.unwater.org/sites/default/files/app/uploads/2018/10/WaterFacts_water_and_watewater_sep2018.pdf (accessed 22 March 2025).

UN Water (2021). World Water Development Report 2021. Valuing water, p. 1. https://www.unwater.org/publications/un-world-water-development-report-2021. (accessed 22 March 2025).

World Economic Forum and World Resources Institute (2023). 25 countries face extremely high water stress, study finds. Aug 25. https://www.weforum.org/agenda/2023/08/countries-extremely-high-water-stress/ 2Oct, 2023 TK. (accessed 22 March 2025).

World Water Week (2023). 6 Key trends from world water week 2023. WaterFront Daily. Aug 24, 2023. https://www.worldwaterweek.org/news/6-key-trends-from-world-water-week-2023. (accessed 22 March 2025).

von Wright, G. H. (1981). Humanismi Elämänasenteena; Translation in Finnish. Original publication in Swedish: Humanismen som livshållning, 1978 (Humanism as a life stance), Otava, Helsinki.

doi: 10.2166/9781789064162_0011

Chapter 2

Water resources and water use in communities: where does it all come from?

'You can make fire as long as you have enough strength to mince woodchips against each other.' (Source unknown).

'You can get water from the ground as long as you have enough strength to dig.' (Source unknown).

2.1 INTRODUCTION

Of all the water on the planet, only 2.5% is fresh and the rest is saline. In fact, the planet Earth perhaps should be known as 'planet water' since over 70% of its surface is under water. Normally, it is only feasible to clean saline or brackish water through desalination for use by communities and industries if fresh water is not available. Such desalination plants are used, for example, in North America, the Middle East, and in the Mediterranean region. In the latter, freshwater supplies are often scarce and water demands are high. Desalination has been adopted for alleviating water scarcity, particularly along coastal areas, where freshwater resources are extremely limited. For economic, environmental and other reasons all existing freshwater cannot, however, be used. Consequently, water quality, water quantity and water equity are prerequisites for society and nature to function (Kattel, 2019).

In many countries, water services (i.e. network-based water supply and wastewater services, and on-site systems and facilities, defined as self-supply systems) are owned by individuals, property owners/custodians (self-supply), various entities or authorities, consumer-managed cooperatives (water users' associations), municipal water utilities, various types of supramunicipal (intermunicipal or regional) organisations/entities, and private enterprises.

The physical water service systems and/or entities and organisations can be connected to each other, or they can cooperate in various individual ways. The World Health Organization recently launched new Guidelines for Drinking Water Quality: Small Water Supplies, which classify three types of water services management – by households, by communities and by professionally managed systems (WHO, 2024).

Since water is used mainly locally, the Dublin Statement on Water and Sustainable Development in 1992 recommended that decisions are taken at the lowest appropriate level, with full public consultation and involvement of users in the planning and implementation of water projects (ICWE, 1992). Furthermore, the overall trend in many parts of the world is towards municipal or local government-owned systems, although institutional arrangements may vary case by case and country by country.

Sustainable water infrastructure is *critical* to ensuring the social, environmental and economic sustainability of the communities served by water utilities (EPA, 2012). This depends, however, on the practices relating to the following three levels that support each other (Figure 2.1):

(i) Sustainable Water Infrastructure: Sustaining collection and distribution systems, treatment plants and other infrastructure that collect, treat and deliver water-related services.

(ii) Sustainable Water Sector Systems: Sustaining all aspects of the utilities and systems that provide water-related services.

(iii) Sustainable Communities: Promoting the role of water services in furthering the broader goals of the community.

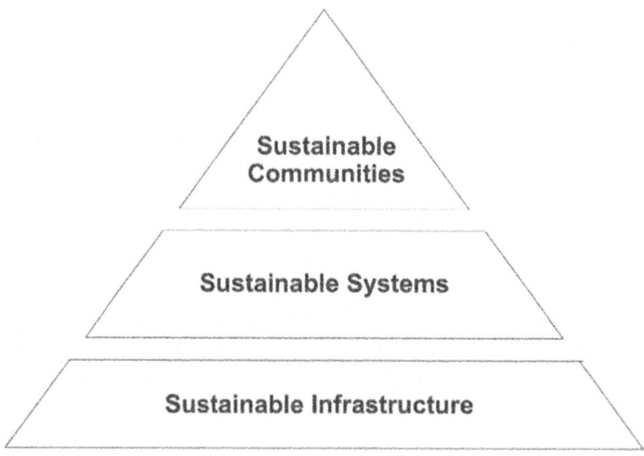

Figure 2.1 Hierarchy of sustainable communities.
Source: EPA (2024).

In this chapter, the focus is on sustainable water infrastructure, exploring the myths connected to water resources occurrence and raw water source selection in communities as well as to the kind of systems that are needed to be able to provide water services for residents and other water users. The framework in Figure 2.1 is further developed for institutional development in the introduction to chapter three.

Myth 1: Water resources and water services mean the same thing

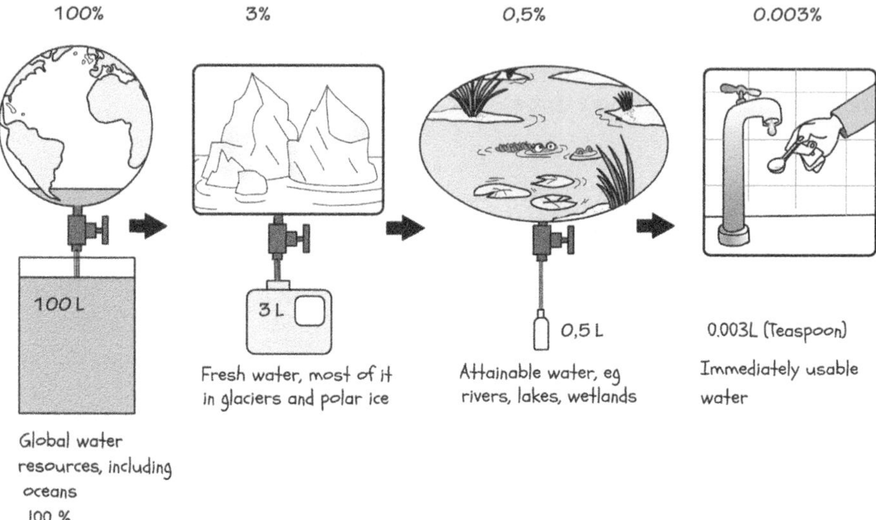

Figure 2.2 Relative shares of salty and fresh water on earth.
Illustration: Kuusisto (1994, modified by Pertti Väyrynen).

In many countries, rivers provide the majority of surface water resources. In some countries, this is from lakes, while others rely exclusively on groundwater. For instance, in Finland there are approximately 168,000 lakes of which 56,000 are over one hectare (Toivonen, 2019). Neighbouring Sweden has even more lakes: 227,000 with an area over 1000 m² (Åberg, 2009). In Norway, communities rely exclusively on surface waters while in Denmark it is groundwater. Geological and geomorphological conditions set the boundaries of what is feasible in terms of abstracting water for community and other uses.

According to Jasechko *et al.* (2024), rapid groundwater level declines are widespread in the 21st century, particularly in dry regions with extensive croplands. These declines have accelerated over the past four decades in 30% of the world's regional groundwater aquifers; this demands special attention be paid to more effective methods of groundwater management.

In many parts of the world, there are strong seasonal variations in rainfall, whereby a major part of the year is subject to drought, with a short season of excess rainfall and flooding. Due to the prohibitive costs of transporting water at least to other continents or across very long distances, it is preferable to try to solve water problems locally as much as possible.

If we imagine the Earth as a 100-litre water barrel, all the fresh water would represent only three litres of this, half a litre would be available for the human race, but only half a teaspoon would be easily accessible fresh water (Figure 2.2). Much of this is groundwater used by over half the world's population. Yet, groundwater deposits are often not known about, and in many cases are overused (Kuusisto, 1994).

According to a recent study by Richardson *et al.* (2023), the Earth has already passed beyond six of the nine planetary boundaries and the latest of them concerns fresh water. This study used streamflow as a proxy to represent blue water (surface and groundwater) and root zone soil moisture to represent green water (plant available water). Water quality related issues were also considered in relation to the other boundaries, namely biogeochemical flows and novel entities.

Other sources contend that in spite of seasonal shortages and scarcity, the world is not facing a water crisis as such, being much more a question of its *lack of proper management.* The World Water Development Report (UN, 2003) concluded that 'The water crisis is essentially about how we as a society perceive and govern water resources and services'. Gleick (2023) reminds us that our economic and political systems have traditionally valued only extraction, monetisation, and consumption of resources rather than efficiency, sustainability and environmental protection. According to him, this is now changing towards healthy environments and ecological restoration.

The lack of clean water is mainly caused by human activities such as water pollution, overuse of water, increased demand for clean water, and mismanagement of water resources. In many regions, water pollution and the lack of action to do something about this are probably the worst culprits.

REALITY: The use, planning and governance of water resources are needed at several levels from international treaties and transboundary waters to water bodies. Water services, for their part, are governed at the local level where citizens use them several times a day. In many countries, the responsibility for providing and developing water services lies with municipalities or other public sector institutions.

Water resources use, planning and governance are needed at several levels from transboundary waters to water courses and waterways of various sizes. Water resource governance is also linked to water services (Figure 2.3). Worldwide, there are approximately 285 transboundary rivers and lake basins that are shared by two or more countries. Altogether, 153 countries have territory within such areas. In addition, there are 592 transboundary aquifer systems (UN Water, 2024). However, it is worth noting that water services and their governance are closer to the end users of water (citizens, institutional users and industries within these communities). According to the second Dublin Principle, they should be managed at the lowest appropriate level (ICWE, 1992):

Water is used for many purposes by communities and societies; their relative shares of this resource vary depending on the specific situation. However, priorities for water use are surprisingly similar around the world in spite of

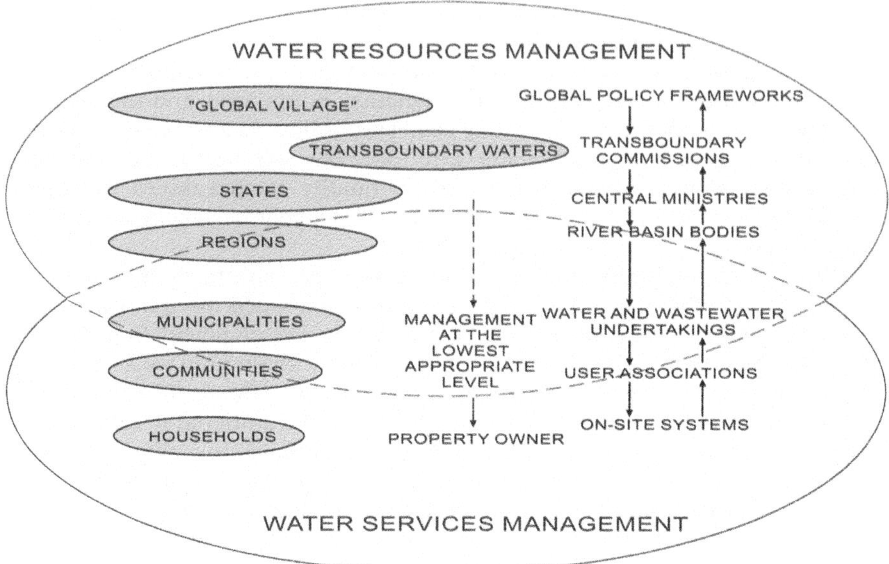

Figure 2.3 Water management: interactions and levels of water resources and water services management.
Source: Pietilä (2006); Katko (2016, p. 19, modified).

differences in the quantities of water used and the diverse socio-economic conditions that prevail. A survey was distributed in 2001 to 2002 to students in ten countries in various continents. They were asked to rank ten uses of water according to first, their actual use, and second, how they should be used. In the latter category, on average, community water supply was ranked first, nature conservation second and hydropower third. Following this were: industrial water supply, irrigation, flood control and drainage, fisheries and fish farming, recipients of wastewater, recreational use, and waterborne transportation. In general, the priorities suggested varied less than was originally anticipated. The survey also implied that the priorities of water use should be a consideration in integrated water resources (Katko & Rajala, 2005).

The International Law Association (2004) stated that water and wastewater services fulfil the 'vital human needs' of communities and, therefore, play a fundamental role in societal and community development. Thus, the governance of water resources includes water quantity, water quality and water use priority in relation to other purposes.

In some countries, even water resources have been privatised such as in Chile where tradable water permits are operated (Water Resources Management in Chile, n.d.). To the researchers' knowledge, in Finland and most other countries, legislation guarantees that water for community purposes has the highest priority. Landowners do not own water resources, including groundwater. However, they have the right to take water for their own use (Vesilaki, 2011).

Related to the priorities of water use is the issue of 'virtual water', which was formerly a very popular concept, referring to water needed to produce a particular quantity of agricultural commodity. However, Warner and Johnson (2007) highlight the fact that 'virtual water' is as state-centric as the 'water wars' thesis. Besides, state security is not the same as household security. Barnes (2013) contends that water fulfils multiple functions beyond watering crops. Furthermore, the concept of virtual water fails to consider the labour needed to channel water onto the fields and the process of the water itself moving through the landscape.

Water service systems lie close to the beneficiaries, that is citizens and communities. In many countries, the duty to provide and arrange water services is increasingly that of local governments (municipalities) or respective public sector authorities. Other major stakeholders often include key ministries, central, regional and municipal authorities, elected officials, water and wastewater utilities, water users' associations, enterprises (consulting companies, contractors and material suppliers), educational and research institutions, professional associations and non-governmental organisations.

For water services, the researchers apply the subsidiarity principle, one of the cornerstones of the Maastrict Treaty, the Charter of the European Union (Maastricht Treaty, 2024). An action is to be taken only if 'by reason of the scale or effects' the objectives cannot be more 'efficiently' achieved by the Member States themselves.

Another important fact is that the global water crisis is a women's issue: in what UNICEF calls 'a colossal waste of time', women and girls spend an estimated 200 million hours hauling water every day (Concern Worldwide USA, 2021). Furthermore, due to the lack of access to clean water and sanitation services, diarrhoea kills 2195 children every day – more than AIDS, malaria and measles combined (Concern Worldwide USA, 2022).

Regarding water services, the distinction between provision and production is important and is discussed in more detail in chapter three.

Connections to water services are one of the most important means of guaranteeing the welfare of citizens and communities. Water and its sustainable use are one major way to eliminate poverty and promote more sustainable habits regarding natural resources. For instance, in Nordic countries and other parts of Europe, existing polluted surface waters have improved to an often good condition thanks to appropriate water pollution control in communities, industries and other point sources. The same is frequently the case for groundwater protection. However, loadings from non-point sources such as agricultural fields, forests, marshland and urban runoff and more recently, brownification of surface waters, are more difficult to control.

Water services currently face and will continue to face major challenges, particularly in areas with high population growth and where populations are declining. In relation to climate change, UNICEF (2023) stated that extreme weather events and changes in water cycle patterns are making it more difficult to access safe drinking water, especially for the most vulnerable children.

Myth 2: Groundwater occurs in veins

Figure 2.4 Traditional groundwater dowsing.
Illustration: Pertti Väyrynen.

Traditional groundwater dowsing (also witching or divining) by a willow rod has been a common practice in Finland and in many other countries. A branch of a willow tree has been used for dowsing so-called 'water veins' in Finland, especially in rural areas (Figure 2.4). In Europe hazel twigs are used, and in the USA this is traditionally witch hazel (Wikipedia.org/Dowsing, n.d.). The assumption has been that groundwater occurs in geological rock formations resembling veins. Some water dowsers use golden rings hanging on a string or a metal rod. Although this tradition has a long heritage, research findings have questioned the efficacy of this belief.

The first written documents on water dowsing date back to the 1500s in Germany, where rods were used in attempting to locate ores (Ellis, 1917). In the 1600s, this practice spread to other parts of Europe and later to the rest of the world (Raittila, 1987). According to Vilkuna (1951), this 'innovation' was diffused to Finland in the late 1800s.

Between 1949 to 1950, the Engineering Department of the National Board of Agriculture in Finland explored the methods used by over 100 well-known water dowsers in the country. They invited 42 recognised experts to locate water 'veins' in the botanical gardens of the University of Helsinki. They

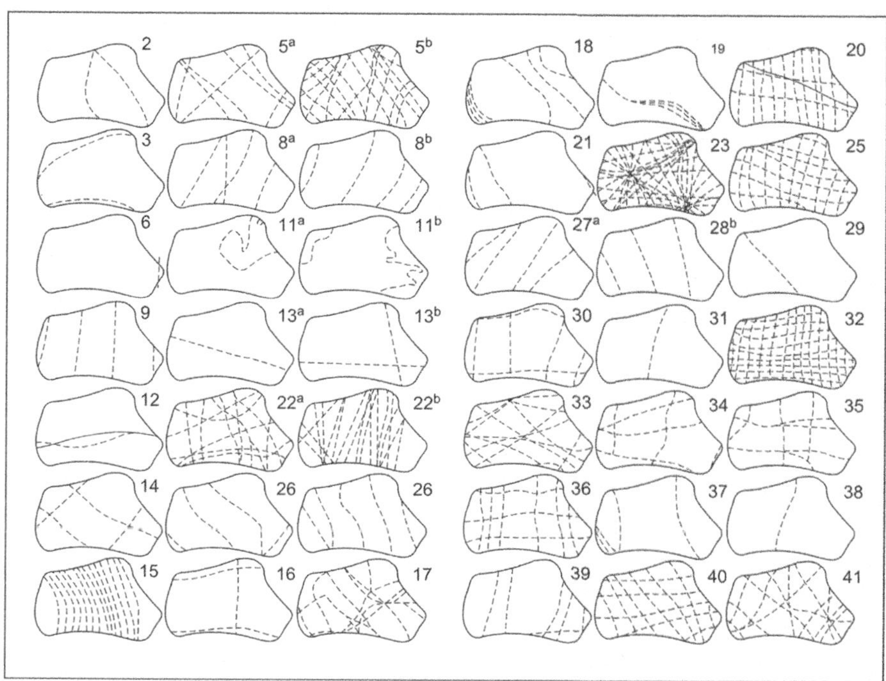

Figure 2.5 Water veins identified by the most famous water dowsers in Finland, 1949–50. *Source:* Wäre (1953, p. 49).

were blindfolded so they could not see the terrain. After the first round, the three most famous dowsers were given another chance, starting from another corner of the area. However, as shown in Figure 2.5, the results were mutually inconsistent. Later studies have come to a similar conclusion in Finland.

In Figure 2.5, the last three results (39–41) of this second trial by the three experts, suggest that the scientific evidence is that dowsing is no more effective than random chance; this is also shown by studies including those in the US in the late 1920s, in Algeria in 1943 to 1944, in New Zealand in 1948, in 1971 in Britain, in 1990 in Munich, Germany, and in 1991 in Kassel, Germany (Wikipedia.org/Dowsing, n.d.).

It is true that the forward pointing tip of a forked willow branch can bend toward the ground when used by a dowser. Often, water seeps through the bottom of the branch, and may give the impression that this comes from a pipe or vein. Yet this phenomenon is not directly linked to the amount, quality or depth of groundwater. Indeed, it is obvious that dowsers are practical 'kitchen geologists' who locate the most potential sites based on landscape and vegetation.

Beliefs in water 'veins' and dowsers remain strong and have spread to many countries such as the USA (Vogt & Hyman, 1959). In a historical context, the researchers found notes on water dowsing, for example from Germany and Spain in the sixteenth century (Wikipedia.org/Dowsing, n.d.). More recently in the early 1980s, a development cooperation project in Sri Lanka, sponsored by Germany, tried to locate water veins with 'modern rods' (Särkioja, 2012). In Namibia in 1994, the Department of Water Affairs announced a training programme for water diviners, again to be sponsored by the German development cooperation (Braune, 2021).

Related to water dowsing is the belief that sensitive people can distinguish water veins by electromagnetic radiation. This was the topic of a doctoral dissertation but it was dismissed by measurement technology in the field of technology and natural sciences (Rahko, 2012). It is probable that this would have been valued more in an ethnologic study.

Colleagues report that this tradition is well known in India (Umare, 2024) and Latin America (Castro, 2024). As an alternative to dowsing, knowledge and interpretations of bedrock geology and geophysical investigations should be used. However, traditional dowsing can still be employed, for example in locating individual wells based on the practical knowledge and experience of dowsers in a specific, small location.

2.2 GEOLOGY MATTERS

Groundwater is water that exists beneath the Earth's surface, filling the spaces between soil particles and rock layers. It primarily resides in the pores and voids found within geological formations known as aquifers (Water Science School, 2019). Groundwater is found in usable quantities most often in sand and gravel formations and in European stalactite caves. An example here is given of a sponge. When wet, it absorbs water into its pores, which can be squeezed out. Similarly, groundwater is held in the pores of materials such as sand, gravel and fractured rock, and it can be extracted through wells (drilled boreholes) (National Groundwater Association, 2023). Various factors influence groundwater movement and availability, including geology, precipitation, human activities and topography.

REALITY: Groundwater occurs practically everywhere in sand and gravel formations. When blind tested, even the best water dowsers produce different maps of the so-called water veins they find. Water dowsers are 'kitchen geologists' who explore and observe nature and use their experience in the trade.

For example, Finland has a unique geological history. Its oldest formations of Precambrian bedrock and its youngest glacial formations have a fundamental impact on groundwater conditions in Scandinavia and Finland (Donner, 1995). The sand and gravel deposits, created during the last ice age, formed the glaciofluvial deposits such as eskers. On the Finnish western coast, acid sulphate soil that was formed during the brackish Litorina Sea period some 7000 years ago, results in a specific quality issue. When these layers are exposed to atmospheric oxygen, for example due to flood control and agricultural activities, particularly high sulphur loadings may occur (Weppling, 1997). Areas with such loadings exist especially in the coastal regions of the Gulf of Bothnia (Figure 2.6). In Kurikka, western Finland, exploitation of the deeply buried groundwater of the valley aquifer system will secure municipal water supply and agro-industrial needs in the future. (Kurikangeologia.fi, n.d.)

The highest elevations of areas with acid sulphate soils range from about +30 m in the south to around +60 m in the west (Figure 2.6). The phenomenon is also connected to the issue of land uplift after the ice age. To use groundwater in any region, it is of fundamental importance to have at least some understanding of the region's geological history, its groundwater occurrence and gravitational flow.

For community water use, a basic question to consider is where water can be drawn from: surface water, groundwater, artificial recharge (also called Managed Aquifer Recharge, MAR, Myth 3) or their combinations (conjunctive use) and how to manage these, especially in growing urban areas?

Figure 2.6 Major areas with known and possible acid sulphate soils in Finland.
Source: Weppling (1997, p. 20, modified).

Myth 3: Managed groundwater aquifer recharge will damage groundwater formations

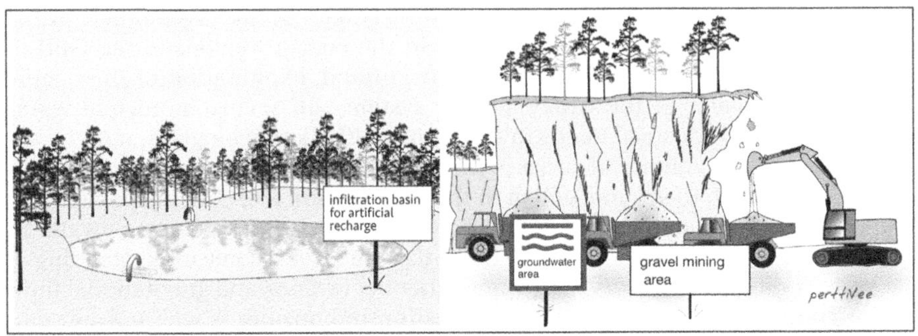

Figure 2.7 Many are worried that artificial groundwater recharge will damage the eskers. Even those who do not seem to be concerned about gravel extraction will soon be close to destroying all existing eskers.
Illustration: Pertti Väyrynen.

For community water services, citizens need to select the most viable of the available options – surface water, groundwater, managed aquifer recharge (previously called artificial recharge), or even desalination of seawater or brackish waters in some cases – or select and combine several of these to formulate the most effective alternative. On a global scale, the use of groundwater has increased continuously, and in some areas has challenged its sustainability, particularly in climatic regions experiencing drought.

Managed aquifer recharge (MAR) refers to the methods used to maintain, enhance and secure groundwater systems under stress. By the 1870s, riverbank filtration for drinking water supplies was well established in Europe. Although the first infiltration basins in Europe appeared in the 1890s in Sweden and France, the more intensive use of MAR began in the 1960s (Dillon *et al.*, 2019).

A fairly common threat scenario, at least among Finnish citizens, is where the infiltrated water taken from a lake or river could spoil, damage or block the esker or the respective sand and gravel formation where water is infiltrated (Figure 2.7). This might cause some environmental impacts depending on the conditions. Yet at the same time there has been surprisingly little concern even by environmental protection bodies about gravel mining where the eskers would be permanently destroyed. The most important groundwater bodies should certainly be reserved for community water supply. Often, groundwater is out of sight and, therefore, it is not widely recognised and understood that this represents up to 98.7% of all freshwater resources (Closas, 2024).

Interestingly, in the 1910s in Finland, there was a strenuous debate on groundwater use and how to estimate its safe and sustainable yield, and more generally, on which raw water to abstract and use, that is surface or groundwater (Katko, 2016, pp. 58–67). More recently, criticisms have come from the general public. Their resistance to MAR projects may be related to a lack of understanding about the nature and extent of groundwater occurrence and its ownership.

In chronological order, Dillon *et al.* (2019, p. 9) recognised five categories of MAR: streambed channel modifications (e.g. infiltration gallery, subsurface dam), bank filtration, water spreading, recharge wells, and runoff harvesting. Previously, MAR was commonly called artificial recharge. The classifications may also vary, for instance in Finland, bank filtration is not officially accounted for under MAR, while the actual options usually cover infiltration basins, sprinkling, or well injection (Katko, 2016, pp. 58–67).

In Europe, Sprenger *et al.* (2017) found a total of 224 MAR systems in 23 countries including Finland, France, Germany, Hungary, the Netherlands, Poland, Slovakia and Switzerland. Through their global MAR inventory, Dillon *et al.* (2019, p. 1) observed that MAR, together with water demand management, is an increasingly important water management strategy, that also protects and improves water quality. However, some specific challenges relating to water quality can still be highlighted. Geological and climatic factors as well as governance structures can have an impact on the application of MAR.

MAR is mainly used for raw water treatment in Finland and Sweden (Kolehmainen, 2008), whereas in the USA, Australia and many European countries it is used for storage of 'surplus' surface water supplies in groundwater aquifers (aquifer storage and recovery, ASR) (Balke & Zhu, 2008; Barnett *et al.*, 2000). In the USA, the number of underground storage systems increased from three in 1983 to 72 in 2005, and this was estimated to be increasing (National Research Council, 2008).

In Oregon state, USA, there have been interesting changes in the use of MAR. Initially, it focused on supplementing drinking water supplies, however, after federal regulations required extensive pretreatment, only municipalities with water treatment plants could afford it. MAR remained in use in areas with undergoing groundwater depletion. Its use was also reestablished in municipalities that conjunctively used surface water and groundwater systems but faced issues with harmful algal blooms in their storage reservoirs. Fires are common in Oregon and regularly threaten surface water supplies with burned materials (Jarvis, 2023).

According to Eberhard and Israel (2021, p. 12), in the African context, MAR is practiced most in South Africa (17 reported cases), followed by Tunisia (11 cases), Kenya (eight cases) and Algeria (five cases). In South Africa, remarkable progress has been achieved on MAR, driven by the research and programmes over the last 50 years of the Water Research Commission. However, the bottleneck in MAR's implementation is not due to a lack of knowledge or technical skills, but is rather caused by slow institutional development (Eberhard & Israel, 2021, pp. 69, 75).

In Finland, community water supply has increasingly relied on natural and artificially recharged groundwater as the raw water source. The first modern MAR systems in Finland came into use in the 1970s (Katko, 2016, p. 63). Several MAR projects have proceeded well due to the cooperation between the parties involved, while some have faced considerable resistance from the general public or some stakeholders.

According to Kokko (2023), the content and objective of the precautionary principle stipulated in the new Nature Conservation Act are no longer coping with the well-established precautionary principle applied in the international biodiversity legal system, in the national judicial culture or in the legal practices of the Supreme Court of Finland and the Court of Justice of the European Union. One major MAR project has already been terminated in 2024, due to the Supreme Administrative Court's recent interpretation and judgment about this controversial and problematic principle. This MAR was assessed by experts to be much more economical and ecofriendly as well as a more reliable and safer water source option compared to surface water.

MAR plants in Finland provide supplies for 16% of the country's total drinking water (Kurki *et al.*, 2013). These systems mostly use basin recharge, sprinkling infiltration and a few use well injection (Katko, 2016, pp. 62–65; Jylha-Ollila *et al.*, 2020). It is often difficult to assess the actual magnitude of filtration since the permeability may change.

The share of groundwater and surface water use for urban areas may vary remarkably depending on the country and region and its hydrogeological conditions. For instance, of the Nordic countries, Denmark uses only groundwater for drinking, industry and agricultural purposes (Jørgensen *et al.*, 2017, p. 1), while in Finland, the use of combined groundwater and MAR artificial recharge accounts for 65% of community use (Rintala, 2020). Sweden provides surface water to 60% of its communities and combined groundwater and MAR to 40% (Statistiska centralbyran, 2017). Norway abstracts only surface water.

Other countries with high ratios of groundwater use include Lithuania, Slovakia and France (IWA, 2012; cited by Jørgensen *et al.*, 2017, p. 5). Globally, groundwater is an essential resource that provides the largest freshwater storage, apart from the ice caps. Current groundwater abstraction represents 26% of total freshwater withdrawal globally, supplying almost half of all drinking water. For drinking water supply, one advantage of groundwater is that it is naturally protected from many contaminants.

The Turku region MAR project is the largest in Finland, currently supplying approximately 65,000 m^3 of water per day. Although it took 40 years to implement the scheme, the delay allowed more time to use and develop modern hydrogeological surveys such as 3D modelling within the MAR area (Artimo *et al.*, 2008). According to Zheng *et al.* (2021, p. 22), 'The system is extensively monitored and managed to ensure compliance with environmental and health regulations'. In this case, thorough hydrogeological and hydrochemical investigations produced the required critical knowledge (Zheng *et al.*, 2021, p. 71).

> **REALITY:** Groundwater can be drawn from and artificially recharged water produced in sand and gravel formations. Managed groundwater aquifer recharge may, to a minor extent, have some harmful impacts on the environment. If needed, water to be infiltrated can be pretreated. However, long-term gravel mining is a much greater threat, and the most important groundwater areas should be protected from this to preserve groundwater use and MAR purposes.

Drawing on a range of sources, the major advantages of MAR systems are commonly seen to include a planned overall system, good and homogeneous quality, an almost stable temperature, and the non-use of chemicals. Disadvantages may cover the risk of raw water pollution, possible problems due to blue-green algae occurring in the surface waters in the warm seasons, impacts on the recharging environment, and potential economic and social contradictions due to competing interests. In case of raw water pollution, it is possible to temporarily shut down raw water abstraction without stopping water supply.

Conjunctive water management allows the combined use of surface and groundwater and other non-conventional sources. For instance, in Geneva, Switzerland, household water is supplied partly from groundwater and partly from Lake Geneva, while the groundwater component is replenished by recharge from the Arve River (de los Cobos, 2002). Groundwater supply to Berlin, Germany, is bank filtrate from percolated stormwater and surface water (IWA) as is the traditional water drawing through riverbanks upstream of the river Danube in Budapest, Hungary (Nagy-Kovács *et al.*, 2019).

Under Finnish conditions, groundwater has at least three advantages over surface water in urban areas. It is often of better quality, its temperature is more stable, and it is better protected from immediate surface contamination. Groundwater use can, however, be limited by excess iron and manganese, particularly in the coastal regions, as well as by excess fluoride in weathered bedrock in some locations. Use of MAR can help mitigate this (Juuti *et al.*, 2023).

The use of groundwater for drinking purposes can have serious limitations. The global distribution of geogenic high arsenic groundwater is mainly in inland basins and river deltas in South Asia, East Asia, and South America (Guo *et al.*, 2024). Small amounts of fluoride are helpful but in excess this may pose a serious health risk (Dar & Kurella, 2023). Additional challenges can be the result of nitrogen pollution of groundwaters (Sun *et al.*, 2024).

The many advantages and disadvantages of surface and groundwater depend largely on local conditions and scale, factors which may also be contradictory. Looking to the future, it is necessary for any country to maintain both water sources in as good condition as possible. This will also improve overall security, and the security of supply for water services. The use of MAR also has future potential for reuse and storage of water resources using nature-based solutions.

Myth 4: Water comes from the tap, and that is all we need

Figure 2.8 Water comes from the tap and electricity from the plug by itself. Or do they? *Illustration:* Pertti Väyrynen.

Water comes from the tap and electricity from the plug. Or do they just appear by themselves? Occasionally, it has been noted that the general public may deem that water services and the systems behind them are simple and easy to manage. While opening the tap, a water user seldom thinks about everything that is needed to have clean piped water, with wastewaters collected and treated efficiently and continuously, 24 hours a day (Figure 2.8).

Overall, planning, construction, and operation and maintenance of water service systems require appropriate technology, economic resources, various types of governance systems, and versatile competences. As a starting point, in developed countries water services are assumed to operate 24 hours a day, seven days a week. This is seen as a goal for all to strive for.

> **REALITY:** Water service systems are complicated networks that are largely invisible and underground. Their construction, operation and maintenance require technology applicable to local conditions, a high level of capital, a variety of governance systems and versatile competencies.

As discussed in Myth 1, integrated water resources management (IWRM) has been internationally promoted rather than individual water-related sectors. This is emphasised in those countries with scarce water resources but which often have significant seasonal rainfall variations and competing water use purposes.

In relation to the most important purpose of freshwater use – water and wastewater services – relatively little attention has been paid to the fact that water and wastewater utilities could in many cases be merged at least at a local level (Juuti & Katko, 2005). In supramunicipal and other larger systems, water service areas may differ, thereby making such consolidations more difficult. Yet, these need to be considered case by case. It is the same water that is abstracted for community use and after efficient treatment is discharged to water bodies. Most often, water is returned to a different location downstream of the water body. If anything is integrated, why not these services?

Stormwater is often included as part of urban water management, although it may be subject to its own legislation. Stormwater management aims to reduce runoff of rainwater or melted snow onto streets, lawns and other sites (EEC Environmental). However, with climate change, and changing rainfall and weather patterns, stormwater management is becoming vital. Another important consideration is the tradition of sewerage management. Older cities in many parts of the world still have a large network of combined sewers. During periods of heavy rainfall, their capacity is exceeded, and wastewaters cannot usually be treated other than by mechanical methods. This can cause major overloads to water bodies and the environment. The direct impacts of this problem were evidenced during the summer Olympics in Paris 2024, when the water quality of the river Seine caused some swimming competitions to be postponed (Bose, 2024). By comparison, newer cities often have separate sewers (Figure 2.9). It may be that in the future, the first flushes of stormwater will need to be treated (Maniquiz-Redillas *et al.*, 2022).

Sustainable stormwater management makes it more important to have proper cooperation and dialogue between land use planning and water services. Although tightening of urban infrastructure and increasing the density of the built environment are a current fashion, this certainly has its limits. Citizens appreciate adequate recreational areas such as parks and the natural environment. To manage stormwater, people need to leave sufficient open, unpaved space as well as retention areas for stormwater routes while preparing for excess rainfall. Furthermore, groundwater protection has to be considered in urban planning. For this reason, the term integrated urban water management is used (IUWM) (Bahri, 2012).

In addition to conventional water sources, there are alternative water supply sources that cities can develop to achieve urban water security. In Singapore, NEWater process recycles treated used water into ultra-clean, high-grade reclaimed water, which supplies up to 40% of the area's water needs. In Melbourne, a recycled water project aims to deliver high quality, class A recycled water to housing estates. Also, a recycled water project is being implemented in two locations in Hamburg, Germany: one is a theme park for educational purposes, the other a planned project for 630 residential units that will separate out blackwater, greywater, and stormwater. In Tucson, Arizona, the utility gives rebates for qualifying rainwater harvesting systems in two categories up to $2000 per property (Brears, 2018). In Australia, Queensland, Urban Utilities supplies 'fit for purpose recycled water' – that is various qualities of recycled water treated to meet the customers' requirements, with the price corresponding to the quality (Brears, 2018). One more option is Managed Aquifer Recharge (MAR) or Artificial Recharge discussed in Myth 3.

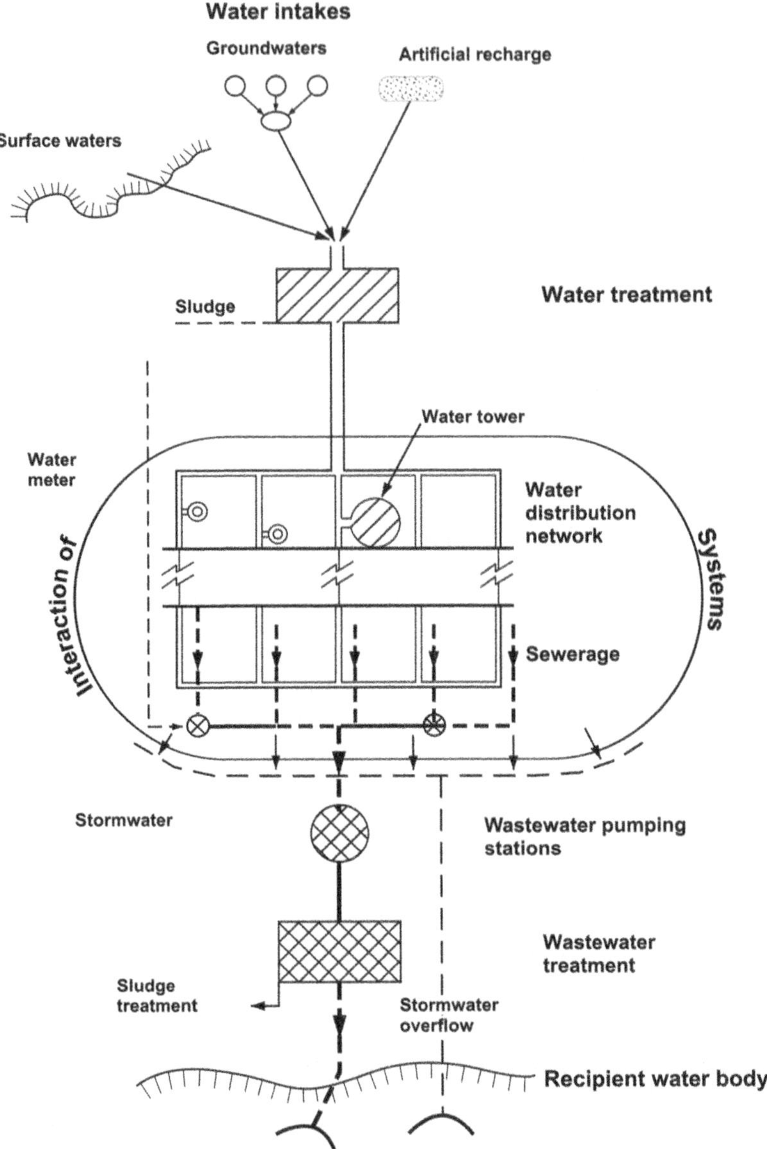

Figure 2.9 Connections of water, wastewater and stormwater systems in a city with separate sewers.
Illustration: T. Katko 2025.

2.3　DISCUSSION QUESTIONS

(1) How do you see water and water services in terms of the SDGs?
(2) What are the pros and cons of using groundwater as raw water for community water supply?
(3) Why do some people perceive managed groundwater aquifer recharge as a greater threat to the environment than gravel mining?
(4) What is needed in order for community water supply to remain in continuous operation, 24 hours a day, seven days a week?

REFERENCES

Åberg J. (2009). Det verkliga antalet sjöar i Sverige – fler än vad SMHI vill räkna? (The real number of lakes in Sweden – more than SMHI want to recognise?), p. 2. http://janaberg.se/wordpress/wp-content/uploads/2009/12/antal_sjoar.pdf (accessed 22 March 2025)

Artimo A., Sarapera S. and Ylander I. (2008). Methods for integrating an extensive geodatabase with 3D modeling and data management tools for the virttaankangas artificial recharge project, Southwestern Finland. *Water Resources Management*, **22**, 1723–1739, https://doi.org/10.1007/s11269-008-9250-z

Bahri A. (2012). Integrated Urban Water Management. Policy Brief. GWP Technical Committee Background Paper No. 16, pp. 5–7. https://www.gwp.org/globalassets/global/toolbox/publications/background-papers/16-integrated-urban-water-management-2012.pdf (accessed 22 March 2025)

Balke K. and Zhu Y. (2008). Natural water purification and water management by artificial groundwater recharge. *Journal of Zhejiang University Science B*, **9**(3), 221–226, https://doi.org/10.1631/jzus.B0710635

Barnes J. (2013). Water, water everywhere but not a drop top drink: the false promise of virtual water. *Critique of Antropology*, **33**(4), 371–389, p. 371, https://doi.org/10.1177/0308275X13499382

Barnett S. R., Howles S. R., Martin R. R. and Serges N. Z. (2000). Aquifer storage and recharge: innovation in water resources management. *Australian Journal of Earth Sciences*, **47**, 13–19, https://doi.org/10.1046/j.1440-0952.2000.00760.x

Bose D. (2024). Here's what to know about Seine River water quality during the Paris Olympics. AP, 30 July 2024. https://apnews.com/article/olympics-2024-seine-water-quality-bacteria-f04236aa61b936912f9af65f7e3fa9ba (accessed 22 March 2025)

Braune E. (2021). A note on Water Divining. 21 July 2021. https://gwd.org.za/news-articles/note-water-divining-eberhard-braune (accessed 22 March 2025)

Brears R. C. (2018). Fit for Purpose Recycled Water in Australia. Medium, 27 Jun 2018. https://medium.com/mark-and-focus/fit-for-purpose-recycled-water-in-australia-d163726cc85b (accessed 22 March 2025)

Castro E. (2024). 28 May 2024, Personal communication, Brazil.

Closas A. (2024). Review of 'Groundwater Sustainability. Conception, Development and Application', by Robert Mace, Palgrave Studies in Environmental Sustainability, Palgrave MacMillan, London, UK, http://www.water-alternatives.org/index.php/boh/item/368-mace (accessed 22 March 2025)

Concern Worldwide USA (2021). Why water is a women's issue. https://concernusa.org/news/water-is-a-womens-issue/ (accessed 22 March 2025)

Concern Worldwide USA (2022). Ten causes of the global water crisis. https://concernusa.org/news/global-water-crisis-causes/ (accessed 22 March 2025)

Dar F. A. and Kurella S. (2023). Fluoride in drinking water: An in-depth analysis of its prevalence, health effects, advances in detection and treatment. *Materials Today: Proceedings*, **102**, 349–360, p. 349. https://doi.org/10.1016/j.matpr.2023.05.645

de los Cobos G. (2002). The aquifer recharge system of geneva, Switzerland: a 20 year successful experience. Management of aquifer recharge for sustainability. In: Proceedings of the 4th International Symposium on Artificial Recharge of Groundwater. P. J. Dillon (ed.) CRC Press, London, Adelaide, London, pp. 49–52. https://www.taylorfrancis.com/books/edit/10.1201/9781003078838/management-aquifer-recharge-sustainability-dillon (accessed 22 March 2025)

Dillon P., Stuyfzand P. [...] and Sapiano M. (2019). Sixty years of global progress in managed aquifer recharge. *Hydrogeology Journal*, **27**:1–30. https://doi.org/10.1007/s10040-018-1841-z

Donner J. (1995). The Quarternary History of Scandinavia. World and Regional Geology 7. Cambridge University Press, Cambridge, UK.

Eberhard B. and Israel S. (2021). Managed Aquifer Recharge: Southern Africa. The Groundwater Project, Guelph, Ontario, Canada. https://doi.org/10.21083/978-1-77470-006-8

EEC Environmental. What Is Stormwater Management and Why Is It Important? https://eecenvironmental.com/what-is-stormwater-management/ (accessed 22 March 2025)

Ellis (1917). The Divining Rod. A History of Water Witching. Water-Supply Paper 416. United States Geological Survey, Washington DC, p. 13.

EPA (2024). Water Infrastructure Challenge. https://www.epa.gov/sustainable-water-infrastructure/water-infrastructure-challenge (accessed 22 March 2025)

EPA (United States Environmental Pollution Agency) (2012). Planning for Sustainability: A Handbook for Water and Wastewater Utilities, prepared by Ross & Associates Environmental Consulting, Seattle, USA, p. 1. https://www.epa.gov/sites/default/files/2016-01/documents/planning-for-sustainability-a-handbook-for-water-and-wastewater-utilities.pdf (accessed 22 March 2025)

Gleick (2023). The Three Ages of Water: Prehistoric Past, Imperiled Present, and A Hope for the Future. Public Affairs/Hachette, New York, p. 296.

Guo J., Cao W., Lang G., Sun Q., Nan T., Li X., Ren Y. and Li Z. (2024). Worldwide distribution, health risk, treatment technology, and development tendency of geogenic high-arsenic groundwater. *Water*, **16**(478), 1. https://www.mdpi.com/2073-4441/16/3/478 (accessed 22 March 2025)

ICWE (International Conference on Water and the Environment (1992). Dublin Statement on Water and Sustainable Development. https://ielrc.org/content/e9209.pdf (accessed 22 March 2025)

International Law Association (2004). The Berlin Rules on Water Resources, Fourth Report, 20. Vital Human Needs, Berlin, Germany, pp. 10, 12, 22–24, 43.

IWA (2012). International Statistics for Water Services. International Water Association, Specialist Group, Statistics and Economics, Montreal, cited by Jørgensen *et al.* 2017.

IWA. Berlin, Germany. https://iwa-network.org/city/berlin/ (accessed 22 March 2025)

Jarvis T. (2023). 18 Sept 2023, Personal communication, Institute for Water and Watersheds, Oregon State University.

Jasechko S., Seybold H., Perrone D., Fan Y., Shamsudduha M., Taylor R. G., Fallatah O. and Kirchner J. W. (2024). Rapid groundwater decline and some cases of recovery in aquifers globally. *Nature*, **625**, 715–721, p. 715, https://doi.org/10.1038/s41586-023-06879-8

Jørgensen L. F., Villholth K. G. and Refsgaard J. C. (2017). Groundwater management and protection in Denmark: a review of pre-conditions, advances and challenges. *International Journal of Water Resources Development*, **33**(6), 868–889.

Juuti P. and Katko T. (eds) (2005). Water, Time and European Cities. History Matters for the Futures. Tampere University Press ePublications, Tampere, Finland, p. 242. https://trepo.tuni.fi/handle/10024/65706 (accessed 22 March 2025)

Juuti P. S., Juuti R. P., Katko T. S., Lipponen A. M. and Luonsi A. A. O. (2023). Groundwater option in raw water source selection and related policy changes in Finland. *Public Works Management & Policy*, **28**(2), 189–214, pp. 193–194. https://doi.org/10.1177/1087724X221100554

Jylha-Ollila M., Laine-Kaulio H., Niinikoski-Fusswinkel P., Leveinen J. and Koivusalo H. (2020). Water quality changes and organic matter removal using natural bank infiltration at a boreal lake in Finland. *Hydrogeology Journal*, **28**, 1343–1357, p. 1343, https://doi.org/10.1007/s10040-020-02127-9

Katko T. S. (2016). Finnish Water Services – Experiences in Global Perspective. Finnish Water Utilities Association, Helsinki, Finland. Co-published E-book, IWA Publishing, London, 2017. www.finnishwaterservices.fi (accessed 15 Dec 2024)

Katko T. S. and Rajala R. P. (2005). Priorities for fresh water use purposes in selected countries with policy implications. *International Journal of Water Resources Development*, **21**(2), 311–323, https://doi.org/10.1080/07900620500108650

Kattel G. R. (2019). State of future water regimes in the world's river basins: balancing the water between society and nature. *Critical Reviews in Environmental Science and Technology*, **49**, 1107–1133. https://doi.org/10.1080/10643389.2019.1579621

Kokko K. (2023). Varovaisuusperiaatteen muutostarve luonnonsuojelulaissa (The need for change of the precautionary principle in the Nature Conservation Act). Edilix, Ympäristö 7 Nov 2023.

Kolehmainen R. (2008). Natural Organic Matter Biodegradation and Microbial Community dynamics in Artificial Groundwater Recharge. Doctoral dissertation no. 781, Tampere University of Technology, Finland, p. 2. https://trepo.tuni.fi/handle/10024/113876 (accessed 22 March 2025)

Kurki V., Lipponen A. and Katko T. (2013). Managed aquifer recharge in community water supply – the Finnish experiences and some international comparisons. *Water International*, **38**(6), 774–789, p. 775. https://doi.org/10.1080/02508060.2013.843374

Kurikangeologia.fi (n.d.). Kurikka's Deep Groundwater Project https://kurikangeologia.fi/en/kurikkas-deep-groundwater/ (accessed 22 March 2025)

Kuusisto E. (1994). Ikuinen, vaan ei muuttumaton. Veden kiertokulku (Eternal but not stabile. Hydrological cycle of water). *Kehitys*, **22**(3), 36–38.

Maastricht Treaty (2024). Edited, 14 August 2024. http://en.wikipedia.org/wiki/Maastricht_Treaty (accessed 22 March 2025)

Maniquiz-Redillas M., Robles M. E., Cruz G., Reyes N. J. and Kim L-H. (2022). First flush stormwater runoff in urban catchments: A bibliometric and comprehensive review. *Hydrology*, **9**(63), 63. https://www.mdpi.com/2306-5338/9/4/63

National Groundwater Association (2023). Information on Earth's Water. National Groundwater Association, Ohio, USA. https://www.ngwa.org/what-is-groundwater/About-groundwater/information-on-earths-water (accessed 22 March 2025)

National Research Council (2008). Prospects for Managed Underground Storage of Recoverable Water. The National Academies Press, Washington DC, p. vii. https://doi.org/10.17226/12057

Nagy-Kovács Z., Davidesz J., Czihat-Mártonné K., Till G., Fleit E. and Grischek T. (2019). Water quality changes during riverbank filtration in Budapest, Hungary. *Water,*

11, 302. https://www.mdpi.com/2073-4441/11/2/302 (accessed 22 March 2025), https://doi.org/10.3390/w11020302

Pietilä P. (2006). Role of municipalities in water services. Doctoral dissertation no 617, CADWES Research Team, Tampere University of Technology, Finland, p. 30. https://trepo.tuni.fi//handle/10024/115085 (accessed 22 March 2025)

Rahko P. (2012). Väitös kaivonkatsomisesta hylättiin (Dissertation on water witching discarded). Kaleva. 17 Oct 2012. https://www.kaleva.fi/vaitos-kaivonkatsomisesta-hylattiin/1872918 (accessed 22 March 2025)

Raittila R. (1987). Kansanomainen vesihuolto Suomessa (Folksy water services in Finland). Master's thesis, Institute of ethnology, University of Jyväskylä, Finland, pp. 33–34.

Richardson K., Steffen W., Lucht W., Bendtsen J., Cornell S., Donges J., Drüke M., Fetzer I., Bala G. [...] and Rockström J. (2023). Earth beyond six of nine planetary boundaries. *Science Advances*, **9**(37), 1. https://www.science.org/doi/10.1126/sciadv.adh2458

Rintala J. (2020). 21 Sept 2020, Personal communication, Finland.

Särkioja A. (2012). 19 March 2012, Personal communication, Finland.

Sprenger C., Hartog N., Hernández M., Vilanova E., Grützmacher G., Scheibler F. and Hannappel S. (2017). Inventory of managed aquifer recharge sites in Europe: historical development, current situation and perspectives. *Hydrogeology Journal*, **25**,1909–1922, p. 1909. https://doi.org/10.1007/s10040-017-1554-8

Statistiska centralbyran (2017). Vattenanvandningen i Sverige 2015. Statistiska Centralbyran, Stockholm, Sweden, p. 16. https://www.scb.se/contentassets/bcb304eb5e154bdf9aad3fbcd063a0d3/mi0902_2015a01_br_miftbr1701.pdf (accessed 22 March 2025)

Sun Q., Lang G., Liu T., Liu Z. and Zheng J. (2024). Health risk analysis of nitrate in groundwater in Shanxi province, China: a case study of the Datong Basin. *J Water Health*, **22**(4), 701–716, p. 701. https://doi.org/10.2166/wh.2024.320

Toivonen J. (2019). Maanmittauslaitos selvitti: Suomi on tuhansien järvien maa (National Land Survey investigated: Finland is a land of thousands of lakes). YLE, 12 July 2019. https://yle.fi/a/3-10875949 (accessed 22 March 2025)

Umare M. (2024). 28 May 2024, Personal communication, India.

UN (2003). Water for People, Water for Life. UN Word Water Development Report 1, p. 383. https://www.eird.org/isdr-biblio/PDF/Water%20for%20People.pdf (accessed 22 March 2025)

UNICEF (2023). Water and the global climate crisis: 10 things you should know. 2 March 2023. https://www.unicef.org/stories/water-and-climate-change-10-things-you-should-know (accessed 22 March 2025)

UN Water (2024). Transboundary waters. https://www.unwater.org/water-facts/transboundary-waters (accessed 22 March 2025)

Vesilaki (Water Act) 4:2, 3 mom. 27 May 2011/587. https://www.finlex.fi/fi/laki/ajantasa/2011/20110587 (accessed 22 March 2025)

Vilkuna J. (1951). Kaivonkatsoja kansanperinteessä (water dowsing in folklore). *Kalevalaseuran Vuosikirja*, **31**, 168–178, p. 169.

Vogt E. Z. and Hyman R. (1959). Water Witching U.S.A. The University of Chicago Press, USA.

Wäre M. (1953). Kaivot ja maaperä (Wells and soil). In: Vesto Oy (ed.). Vesihuolto-Opas. Vesiteknillinen insinööritoimisto Oy, Helsinki, Finland, pp. 47–51.

Warner J. F and Johnson C. L. (2007). 'virtual water' – real people: useful concept or prescriptive tool? *Water International*, **32**(1), 63–77, p. 74, https://doi.org/10.1080/02508060708691965

Water resources management in Chile (n.d.). https://en.wikipedia.org/wiki/Water_
resources_management_in_Chile (accessed 22 March 2025)

Water Science School (2019). Aquifers and Groundwater. US Geological Survey,
Virginia, USA. https://www.usgs.gov/search?keywords=Aquifers+and+Groundw
ater (accessed 22 March 2025)

Weppling K. (1997). On the assessment of feasible liming strategies for acid sulphate
waters in Finland. Dissertationes, Geographicae Universitatis Tartuensis 5, Tartu,
Estonia, p. 20.

WHO (World Health Organization) (2024). Guidelines for Drinking Water Quality:
Small Water Supplies. WHO, Geneva, Switzerland, p. xviii. https://www.who.int/
publications/i/item/9789240088740 (accessed 22 March 2025)

Wikipedia.org (n.d.). Dowsing, Dated 29 Dec 2024. https://en.wikipedia.org/wiki/
Dowsing (accessed 22 March 2025)

Zheng Y., Ross A., Villholth K. and Dillon P. (eds) (2021). Managing Aquifer Recharge. A
Showcase for Resilience and Sustainability. UNESCO, Paris, France. https://www.
water-reuse-europe.org/managing-aquifer-recharge-a-showcase-for-resilience-and-
sustainability/ (accessed 22 March 2025)

doi: 10.2166/9781789064162_0035

Chapter 3

Water services as a community (blood) circulatory system

'I believe that water is the only drink for a wise man.' (Henry David Thoreau)

'Water is life's matter and matrix, mother and medium. There is no life without water.' (Albert Szent-Gyorgyi)

3.1 INTRODUCTION

Well-functioning water utilities play an important role in societal progress in communities worldwide. It is likely that the benefits of safe and reliable water services are systematically underestimated due to many non-economic advantages, which are difficult to quantify but are of high value to the users in terms of dignity, convenience, social status, cleanliness, health and overall wellbeing. In addition, safe and reliable water services appear to be a key driver for economic growth – including investments by businesses that are reliant on sustainable water services for their production processes and workers (OECD, 2011).

As mentioned in the introduction to chapter two, the three levels of sustainable water services that support each other are: Sustainable Water Infrastructure, Sustainable Water Sector Systems, and Sustainable Communities (EPA 2023). Takala (2017, p. 511) claimed that sustainable development has meaning in relation to water services only if they are examined as part of the community or municipality they serve. Therefore, Takala (2017, p. 503) suggested that sustainable development of water systems needs to be examined as part of public services and the public good. Public good should not, however, be defined by the experts alone but necessitates dialogue and community involvement.

The Nobel Laureate Ostrom (2010) lists several design principles that can be applied to characterise robust institutions for effectively governing and managing public water services (Box 3.1). These design principles appear to synthesise core factors affecting the probability of long-term survival of an institution developed by water service users.

Box 3.1 Institutions for governing and managing effectively public water services

(a) Water service users are authorised to participate in public decision making and modifying related governance rules.
(b) The use and condition of public water services must be monitored.
(c) Effective and efficient conflict resolving mechanism(s) must be introduced.
(d) Governance rules are congruent (in line) with local social and environmental conditions.
(e) Self-governance of local water service users is recognised by higher level authorities.
(f) Comprehensive public water service governance must be organised in multiple levels, where the lowest levels are representing the local community (Ostrom, 2010).

Hooghe and Marks (2002, p. 8) considered that beyond the common agreement that efficient governance must be multi-level, there is no consensus about how this should be structured. However, according to Oakerson (1999), Ostrom and Ostrom (1991) and the Advisory Commission on Intergovernmental Relations (ACIR, 1987), a public service industry or local public economy should have three components. They are jurisdictions and organisations that *provide*, jurisdictions and organisations that *produce* or supply goods or services, and jurisdictions and organisations that *legislate and administer* rules governing provision and production. Heikkila (2004, pp. 102–103) argued that when provision, production and governance are separated, jurisdictions have opportunities to address a shared problem related to the management or use of a resource by coordinating service production. Therefore, interjurisdictional coordination can allow small-scale jurisdictions to engage in conjunctive water management in a way that eliminates the need for a large and centralised authority.

According to Parks and Oakerson (2000, p. 170), local public economies research makes an important distinction between provision of goods and services and their production, as shown in Box 3.2.

Box 3.2 Provision and production
Provision – Decisions made through collective-choice mechanisms:

(1) The type of goods and services to be provided by a designated group of people.
(2) The quantity and quality of the goods and services to be provided.

> (3) The degree to which private activities, related to these goods and
>   services, are to be regulated.
> (4) How the production of these goods and services is to be arranged.
> (5) How the provision and production of these goods and services is
>   to be financed.
> (6) How the performance of the producers of these goods and services
>   is to be monitored.
>
> **Production** – Production refers to the actual process of producing and
> delivering the services on which the provision authority has decided:
>
> (1) The more technical process of transforming inputs to outputs –
>   making a product, or in many cases, rendering a service.
> (ACIR (1987) cited by Ostrom *et al.* (1993)).

Regulatory governance refers to the following:

(a) the application of binding rules – including standard setting, monitoring
     and sanctioning – by which governments can intervene in the activities
     of specific categories of economic, political or social actors to effectively
     regulate society and economy; and
(b) the systematic implementation and operation of government-wide
     policies on how to use regulatory powers to produce quality regulation
     within the procedural values of the governing system (such as democratic
     processes).

The goal of good regulatory governance is to ensure that regulations efficiently
produce economic, social and environmental benefits. This means that widely
defined benefits should justify widely defined costs, they should be the minimum
needed to produce any level of benefits, and resources should be allocated to
their highest values. (Jacobs & Ladegaard, 2010).

Based on the aforesaid three components, Katko and Hukka (2015) proposed
the following multi-level governance pyramid for the sustainable water services
of European Union member countries (Figure 3.1):

According to Heikkila (2004, p. 111), local public economy (or public service
industry) theory conceptualises institutional boundaries as being formed
around local problems that affect a group of individuals, rather than around a
physical resource or good. Low *et al.* (2003) argued that the presence of larger,
overlapping jurisdictions is an important complement to the work undertaken
by parallel, smaller scale units. Larger units can back up the smaller units in
several ways: (i) providing support at times of natural disasters; (ii) addressing
corruption or gross inefficiency; (iii) providing scientific and technical skills to
complement the local knowledge; (iv) providing conflict resolution arenas for
conflicts among parallel units; and (v) taking on functions that are generally
more efficiently undertaken by larger units.

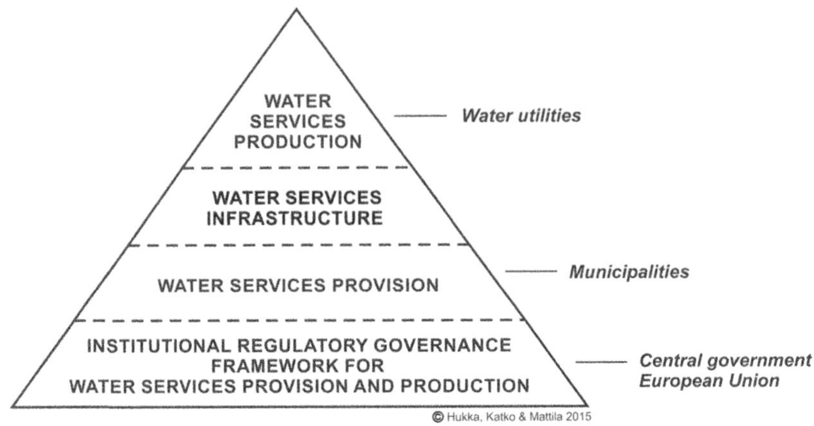

Figure 3.1 Proposed hierarchical framework for sustained water services regulatory governance, provision, infrastructure and services production.
Source: Katko and Hukka (2015, p. 221) modified by the authors (2024).

Multi-level governance planning can be conducted in accordance with the principle of subsidiarity (Papunen (1986), cited by Haapalainen (1996)). Haapalainen (1996) pointed out that this principle has two main dimensions: centralisation and decentralisation. The focus of multi-level planning is on the system, having multiple levels and objectives. It is based on two-way communication between the planning levels and on the expression of their own development interests and objectives. Multi-level planning and the subsidiarity principle can thus be applied to the public policy planning of societal systems, either to promote centralisation or decentralisation.

The issue of balance between a central power and democratic decentralisation and the role of citizens is elaborated by Figure 3.2. Although the conditions in high-, middle- and low-income societies differ, they all face the same fundamental questions:

- Which direction should we move in?
- Since strongly centralised bodies largely failed after decades of trials in developing economies, could regional and local bodies produce better results? History shows this to be the case, at least in Europe.
- What should the balance between centralisation and decentralisation be in each case?

The principle of subsidiarity has emerged to become a pillar of integrated water resources management. As long as subsidiarity is pursued with rigorous and transparent intent, and not as a panacea, water resources and human communities will benefit. The subsidiarity principle of water resources management suggests that water management and public service delivery should take place at the lowest appropriate governance level (Stoa, 2014, p. 45). In many cases, the subsidiarity principle of water resources management has

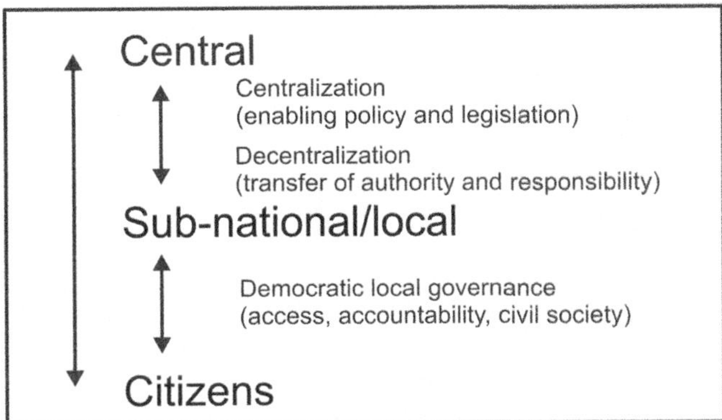

Figure 3.2 Centralization vs. decentralization.
Source: Barnett *et al.* (1997) modified by the authors (2024).

been applied with undue haste, assuming that water resources management should occur at the local level when, in fact, institutional capacities would suggest that local institutions are not the appropriate governance level (Stoa, 2014, p. 32). Chaplin (2014) argued that subsidiarity is better understood as apolitical with respect to how the state should be structured: 'Subsidiarity is a call for social functions to be fulfilled, not at the lowest possible level but rather at the right level'.

Burbridge (2017) believed that a *virtue ethicist* approach meets the demands of many, if not all points raised in discussions of subsidiarity. An advocate of virtue ethics can assert, with regard to problem solving, that there are no common solutions to moral conflicts, but a virtuous personality facilitates making the best possible context-specific decision (The Helsinki Term Bank for the Arts and Science, 2021).

Following this introduction, the respective authors of the book will 'dive into' Myths five to 12, related to water services as a community (blood) circulatory system.

Myth 5: Water treatment and wastewater treatment are expensive

Figure 3.3 A variety of analytics are used in the treatment and control of water and wastewater.
Illustration: Pertti Väyrynen.

A common and understandable belief *per se* is that water treatment and wastewater treatment are expensive (Figure 3.3). Yet, contrary to what is often imagined, for community water service systems, these account for only a small fraction of the cost of water utilities. Both treatment systems combined account for approximately 15–20% of the operating costs (operating expenditures and capital costs) of a water utility. The largest share, approximately 80–85% of operating costs, covers water distribution, sewerage and stormwater networks. However, depending on conditions, advanced wastewater sludge treatment and disposal may cause significant additional costs.

At the same time, about 80% of the costs of water and wastewater management systems are fixed, that is, they are caused by the construction and operational management (operations and maintenance) of the systems, and do not depend on the volume of water used. In many countries, of the total water and wastewater service costs, clean water provision (water intake, treatment and distribution) accounts for approximately 40% with the remaining 60% funding wastewater (sewerage and treatment).

In fact, the majority of water supply needs to be treated to more or less drinking water quality standard; this includes water for cooking, dishwashing, laundry and personal hygiene. In practice, it is only toilet flushing water that can be of lower quality. In Finland this accounts for approximately 25% of household water use. This percentage can be higher in countries where potable water is used for watering lawns and other gardening purposes.

One of the most prevalent myths surrounding water services is the belief that achieving required water treatment quality standards is inherently expensive. The same is true for wastewater treatment. This finding is based on many presentations to various types of citizen groups in different countries. As such, this is understandable: many believe that it is always not sensible to flush faeces and urine using potable water.

A cubic metre of water, including sanitation, costs approximately 2% of the average household net income for a family in Finland. Clearly, water is very cheap compared to any other product. In Europe, in 2021, the mean annual billing rate per person excluding VAT (Value Added Tax) was slightly under 200 euros (with a minimum of 70 and maximum of 320 euros) (EurEau, 2021, p. 6).

However, it is important to exercise caution in comparing water prices between countries and to consider the different elements of tariff structures to make actual water bills comparable. For instance, the percentage of VAT may vary while in some countries this tax does not exist at all. Interestingly, while the average rate of specific water use (litres per capita per day) in the USA is approximately twice that of Europe, average European water bills are double those charged in the USA. This makes the total price the same in both (Hukka & Katko, 2015a, 2015b).

A historic review of water pricing in the USA was conducted by Murdoch Jr. (1956). He concluded that traditionally, water has been too cheap. Such low rates and poor water works have meant that customers have not been able to get the service when they want it. A more recent survey of the annual reports of 14 large water and wastewater utilities in Finland revealed contradictory language used by the owner municipalities on water services: cheap water prices are considered to be a competitive edge while at the same time, necessary investments for ageing infrastructure should be increased (Heino & Takala, 2013).

A report on the State of the world's drinking water (WHO *et al.*, 2022, p. 8) called for action to accelerate progress on ensuring safe drinking water for all. Despite the benefits of safe drinking water, still an estimated two billion people (one quarter of the world's population), lack access to this service. Unsafe drinking water is a major cause of death from diarrhoea for 1.5 million people every year, most of them infants and small children.

However, progress made on increasing access to drinking water is fragile. Safe water is not a one-off investment but requires monitoring and upkeep. Furthermore, external factors threaten these gains: climate change drives water scarcity, flooding disturbs supplies, and water pollution threatens human health, ecosystems and ecosystem services. Urbanisation and population growth limits the ability of cities to deliver water to millions of people in informal settlements, while in rural areas, people struggle with low quality services, waterpoint breakdown, and contaminated water sources (WHO *et al.*, 2022, p. 8).

In more advanced conditions, people tend to take safe water for granted as long as the systems operate without any troublesome disturbances. The cost and price of water is a question that people are interested in, at least in regard to rising water prices. A specialist and associate professor of cooperatives, Dr. Köppä pointed out that the model of sustainable development should consider the impact on the environment and human rights, and not just the cost of production measured in monetary terms. According to him, in the interpretation of economic development, competition almost completely displaces cooperation. To understand economic cooperation, the common reality of human communities must be taken into account. Collaboration is as important as competitiveness. Therefore, the vitality of the totality is more valuable than narrow productivity (Köppä, 2022).

Hukka and Katko (2015a) proposed a wider framework of cost recovery that has five key elements: awareness of benefits and costs, informed policymaking, consumer contributions, tariffs, and fee collection and financial management. Underpricing of water services and the need for increased rehabilitation seem to be worldwide phenomena. Therefore, sustainable water services require users to bear all or at least a major part of the costs. Better awareness of economic and social benefits of water supply, and especially of sanitation, is also needed.

In low- and middle-income economies, the issue of affordability is particularly relevant; often this suggests the use of on-site systems and less complex and sophisticated technologies that people can afford. However, in principle everyone should pay, but not in the same way, by the same amount or at the same time (Laugeri, 1987).

According to the Global Handwashing Partnership (2022), several studies have evaluated the effectiveness of hand hygiene practices in preventing or controlling diseases among children. A meta-analysis by Wolf *et al.* (2022) showed evidence of the effectiveness of handwashing with soap in improving drinking water, sanitation, and the risk of diarrheal disease in children in low- and middle-income conditions, which for the under-fives is reduced by 30%.

REALITY: Out of the overall water service costs, the share of water and wastewater treatment costs is a maximum of 20%. Over 80% of the costs go to water and wastewater pipelines. For economic and environmental reasons therefore, the dual system may be feasible only in certain areas experiencing water scarcity but not in those where adequate water resources are available.

3.2 WHAT ABOUT DUAL SYSTEMS?

In theory, it may sound a nice idea to have dual systems everywhere – separate pipelines for high quality drinking water and other pipelines with lower quality water – rather than wasting high quality water for flushing wastes. However, there are several limitations to such systems – a major one being the costs involved. Contrary to the common assumption, water treatment is relatively cheap while approximately 80–95% of the operating costs are for the networks as discussed earlier. Nevertheless, in the case of water scarcity dual systems may be considered.

In 2011, there were approximately 340 dual systems in the USA, located in the more arid regions of Arizona, California, Florida, and Texas, and serving industry, agriculture or other use (Grigg *et al.*, 2011; Katko, 2013). An exotic dual system is found in Windhoek, Namibia. In 1968, a system was first implemented, where treated wastewater is partially used to provide the raw water for the city's water supply system. The plant has operated well, requiring quite complicated treatment technologies, holistic planning, training and continuity management (Haarhoff & Van der Merwe, 1996). In 2002, the old recycling plant was replaced by one which supplies about one third of Windhoek's potable water (Clayton, 2023). In South Africa, it was noted that the level of aridity of an area is a major driver for non-potable water reuse and the implementation of dual systems (Ilemobade *et al.*, 2009).

In the 1990s, a few dual systems were constructed in Finland for a limited area with a nearby raw water source providing reasonably good water quality. These dual systems were used for laundry and flushing wastes. Previously, it had been possible to cover part of the operating costs through consumer charges, but along with the ageing networks, they needed to be rehabilitated (Kallioinen, 2000). Furthermore, managing such systems also poses institutional challenges and requires adequate resources.

This example is a reminder that in fashionable circular economy projects one should not go forward merely with technical artefacts but should also consider institutional possibilities and limitations: that is, how to manage such systems on sustainable basis? (chapter four) For summer homes or holiday cottages with on-site systems, drinking water can be organised by transporting tap water, for example, in jerrycans or plastic containers. Surface water or groundwater may be used for other purposes while bearing in mind the need for proper treatment and disposal of grey waters (wastewater from baths and showers, handbasins and kitchen sinks, but excluding wastewater and excreta from water closets).

Regarding water quality, reference should be made to Myth 3 (Managed groundwater aquifer recharge will damage groundwater formations) and the major properties of surface water and groundwater that are subject to local conditions. A particular water quality problem may be caused by lead pipes (Myth 10) that are still in use in many older cities.

Myth 6: Bottled water is better than tap water

Figure 3.4 Bottle or tap water? How much does it cost?
Illustration: Pertti Väyrynen.

The sales of bottled and otherwise packaged drinking water have increased in many countries (Figure 3.4). According to a United Nations University (UNU) report based on literature and data from 109 countries, in five decades bottled water has developed into 'a major and essentially standalone economic sector', with an increase in use to 73% from 2010 to 2020. In addition, sales are expected to almost double by 2030 (UNU, 2023, p. 5). It is estimated that less than half of what the world pays for bottled water annually would be sufficient to ensure clean tap water access for hundreds of millions of people who are currently without it – for several years (UNU, 2023, p. 2).

In the Global North, bottled water is often perceived to be a healthier and tastier product than tap water, while in the Global South, sales are driven by the lack of reliable public water supplies largely due to rapid urbanisation. In middle- and low-income countries, bottled water use is linked to poor tap water quality and often unreliable public water supplies (UNU, 2023, p. 10).

In 2023, the worldwide bottled water market generated a revenue of approximately 340 billion US dollars. Compared to previous years, this was a big fall, probably due to the coronavirus pandemic and the restrictions and lockdowns that came with it (Ridder, 2023). This amount is almost three times that needed to meet the UN targets on water (Gleick, 2023, p. 140).

The bottled water market is fuelled by fashion, the ease of use of packaged water, and huge marketing campaigns. In 2018, the highest per capita use of bottled water was found in Mexico, Thailand, Italy, the USA and France (Rodwan Jr., 2019). Tourism increases the use of bottled water and also water and marine pollution (Digap, 2021).

The UN's SDG6 is 'To guarantee safe, equitable and affordable drinking water for all' (UN, n.d.). Large-scale use of packaged water may threaten the achievement of this goal. If the use of packaged water replaces the necessary investments in water supply infrastructure or water abstraction is not organised in an otherwise sustainable way, the situation could lead to further inequality at worst and erode even the smallest water supply systems (Makkonen, 2020).

Another issue related to bottled water is its quality. The Finnish Institute for Health and Welfare (THL) studied the microbiological quality of both spring water vending machines and bottled water, finding that packaged water was rich in cultured bacteria. Bottled waters are not sterile, and microbial growth occurs during storage. The numbers of living microbes were particularly high in samples taken from spring water vending machines (Miettinen & Pursiainen, 2009).

In 2018, bottled water and its microplastics were studied in 19 locations in nine different countries (Walker, 2018). Microplastics over 0.1 mm in size were detected in bottled water; this was twice as much as in tap water. The findings suggested that less than a fifth of the plastic came from the water itself. Although the amounts of plastic being studied are vanishingly small, making it difficult to identify them or assess their effects, it is known that the smallest microplastics and even smaller nano plastics may be harmful to humans (Mason *et al.*, 2018, p. 1; Walker, 2018). Therefore, caution is needed. Another related example is the case of per- and polyfluoroalkyl substances (PFAS) that have been observed to contaminate groundwater, surface water and soil. Cleaning up such polluted sites is technically difficult and costly, while requirements for their treatment is in the process (EPCA, n.d.).

Based on guidelines for international tourists, there are 29 countries in Europe where tap water is safe to drink (Tebbutt, 2016). In the USA, controversy over the use of bottled water in public spaces has continued for many years. In 2014, San Francisco became the first major American city to try to limit the sale of plastic water bottles due to the environmental damage they cause. In 2019, San Francisco Airport banned their sale but did allow water to be sold in glass bottles and recyclable aluminium cans (Clayton, 2019). In 2021, this plastic-free policy was expanded to prohibit the sale of all plastic bottles (Yakel, 2021).

Between 2005 and 2018, Rosinger (2022) noted an increase in bottled water uptake in the USA. In 2017–18, 52% of adults did not drink tap water on a given day, with 36% exclusively consuming bottled water.

REALITY: Bottled water quality tends to deteriorate once bottled. In many countries, tap water quality is better than that of bottled water. In low- and middle-income countries, however, the use of bottled water may be justified to avoid health risks.

Campaigns to defend tap water are ongoing around the world. According to Parag and Roberts (2009), the environmental load from bottled water was more than a hundred times that of tap water. They argue that measures are needed to restore confidence in tap water, such as encouraging proactive water users' participation to understand potential pitfalls, improve transparency, enhance science communication, and increase labelling of environmental hazards in bottles.

In 2017, as part of the centennial events to mark the country's independence, the Finnish Water Utility Association (VVY) organised the Drinking Tap Water Campaign, repeating this in 2024. The World Championships in Athletics for the Under-20s took place at the Ratina Stadium in Tampere, Finland from 10th to 15th July 2018. At that time, the local water utility, Tampere Water supplied the public and athletes with tap water and refillable bottles, which received a lot of positive international attention (Yle News, 2018).

According to an international survey, 97% of Finns trust the quality of tap water in their own homes. By comparison, in countries such as Spain, the United Kingdom, Poland, Germany and the USA, confidence was in the range of 60–72%. The same study showed that Finnish men estimated their monthly water use to be significantly higher than actual use, while Finnish women underestimated their water use (Kemira, 2020).

Use of bottled water is, however, not unambiguous and one can also call into question the common way of comparing bottled with tap water. Bottled water is not necessarily a substitute for tap water, but rather, for other soft drinks. Free public sources of drinking water exist in numerous countries. Thus, bottled water has its pros and cons (ProCon.org, 2021).

In many low- and middle-income countries, piped domestic water is not safe to drink. For instance, in Indonesia, drinking tap water has never been accepted as the norm since piped water has no guarantees of purity and safety. Despite the many negative social and environmental aspects of bottled water, it is becoming 'the' drinking water in Indonesia (Prasetiawan *et al.*, 2017). In Mexico, it was estimated that in 2010 approximately 60% of the population relied on bottled water, with this figure rising by 2017 to more than 70%. This business attracts large-scale investment by multinational food and beverage corporations as well as small-scale neighbourhood water shops (Greene, 2018).

Both globally and locally, the poor suffer from a wide range of illnesses and disadvantages due to the lack of safe water that is often costly. Citing Gleick (2023), O'Riordan (2024) refers to the spread of bottled water as the last bastion of access to clean water for the very poor, who are least able to afford single-use plastic containers.

Packaged water will probably defend its place in water supply, at least in situations and locations where safe tap water is not available. Competition between tap water and packaged water will undoubtedly continue.

Myth 7: Water towers are outdated

Figure 3.5 Smile on your face, open space. Eagle Wisconsin Water Tower (2013, USA). *Illustration:* Pertti Väyrynen.

In many communities, ground-based reservoirs situated at high elevations, or water towers in locations where this is not possible, represent almost the only visible signs of water service systems. In some countries, elevated water reservoirs or water towers have been abandoned or replaced by lower elevation ground or underground reservoirs pumping directly to the network.

In Finland (Nagler, 1966), Sweden and France there is a strong tradition of concrete water towers made by slipform casting, which in practice makes them look like mushrooms (Asola, 2001, pp. 47–49). This tradition and the variety of designs have been noted in international literature. On the contrary, in the USA, towers made of steel are exclusively used (Figure 3.5). Many early water towers are considered to be historically significant and are included in heritage listings around the world (Water tower, n.d.).

In Finland, Asola (2001, pp. 165–173) identified 448 elevated reservoirs completed between 1876 and 2000, of which approximately 400 were still in use in 2001. The boom in water tower construction occurred during the 1960s and 1970s. This can be compared to the number of Finnish municipalities, currently slightly over 300. For very small systems, however, it is too expensive to construct water towers, at least if concrete is used. Larger and city systems may have several, often having different pressure levels as appropriate.

According to Wolf (2020), water towers serve two major purposes: they harness just sufficient gravity to provide adequate water pressure for potable water use, avoiding excess leakages in the network, and possibly serve as a beacon of civic pride.

The storage volume of water towers in Finnish communities varies from 10 to 60% of the average daily water use in them. In 2006, the average total volume of water towers and elevated reservoirs was approximately 45% of their average daily use (Katko & Juuti, 2007). Thus, in case of a major disruption to water reticulation or distribution, they can supply water for a couple of hours. This is particularly important for such critical institutional uses as hospitals. In reviewing critical health infrastructure and its robustness and reliability, Sänger *et al.* (2021) suggested that only rarely are emergency situations exclusively technical in their nature but are more likely to be linked to natural hazards, technical mishaps, and insufficient social and organisational preparedness and resilience.

REALITY: By using water towers together with modern pumping technologies, we can save and optimise energy use. During delivery cuts and shortages, water towers serve as backup units. Incidentally, water towers make water supply more visible to the general public.

Water towers are often the only visible parts of community water service infrastructure. The rest is otherwise largely invisible. Each water tower, in the same way as a building, reflects their own construction styles and traditions (Asola, 2001, pp. 50–141), which may be recognised if one is paying attention to them. Their design can sometimes also reflect relevant aspects of the surrounding environment such as the water tower in Rovaniemi, Lapland, which resembles the shape of a traditional Sami dwelling. Occasionally, an architectural competition has been held to design water towers, although it has often proved quite challenging to find a good balance between economic feasibility and aesthetic requirements (Katko, 2016, p. 96).

Figure 3.6 presents the changes over the years in volume, pressure levels, and design towards mushroom type towers in Tampere, Finland. At first, ground-based reservoirs were constructed followed by actual water towers. With time the volume of water towers has increased as well as their elevations in the suburbs.

In the early days, water towers played a crucial role in providing fire protection by supplying water for firefighting, a function they still serve to some extent today. The Great Fire of Turku in Finland in 1827 is still the largest urban fire in the history of Finland and the Nordic Countries (Salmi, 2017). When water utilities were planned in the late 1800s, fire protection was the first motivation for building the systems. Other cities in Central Europe with predominantly wooden buildings were also commonly subjected to fires, such as the great fires in Berlin in 1727 (Bärthel, 1997) and Hamburg in 1842 (History Timelines).

According to Wolf (2020), there are approximately 70 water towers listed on the National Register of Historic Places in the USA, such as the ornamental detailing and complex symmetry of the Chicago Water Tower (2014), built in 1869, or the simple elegance of the Louisville Water Tower (2024), Kentucky, built in 1860.

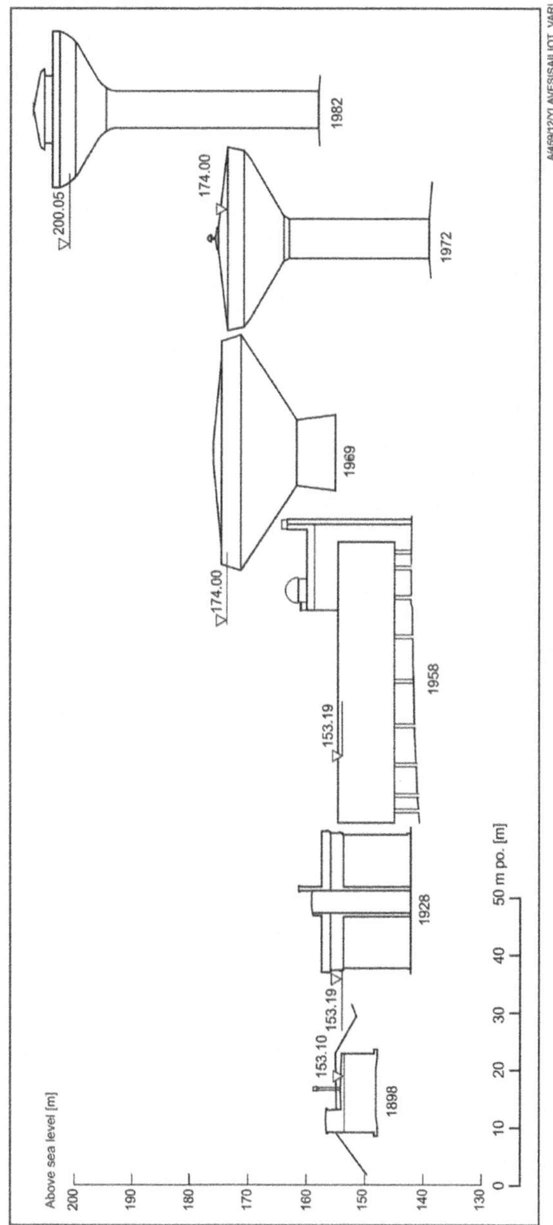

Figure 3.6 Silhouettes of elevated water reservoirs and towers in Tampere, Finland: from left: the 'cellar' on Pyynikki ridge from 1898, the old 'crown' of Kauppi from 1928, the new 'ice-hockey puck' of Kauppi from 1958, and the mushrooms of Tesoma from 1969, Peltolammi from 1972, and Hervanta from 1982. *Illustration:* Katko and Juuti (2007, p. 28).

Sunela and Puust (2017) noted that remarkable savings in energy and use of chemicals can be achieved by using almost real-time optimisation in water treatment and distribution. The model calculates the best possible flow and pressure settings for water treatment plants as well as pumping and valve stations within the network. It takes into account real and estimated water uses, water production rates, pumping capacities and limitations, and tends to use water reservoirs as efficiently as possible.

It has been estimated that in Finland water towers account for one-fifth of the renovation backlog of water services infrastructure. Water tower renovations have been actively on the agenda in the last few decades. Especially during power cuts due to storms or distribution cuts due to pipeline breaks, their storage capacity has proved to be important (Katko, 2016, p. 96).

The question of protection of old buildings and cultural heritage also has relevance to water towers, which often may be among the oldest buildings and landmarks at the most visible sites such as in Obenhausen in the Ruhr area, Germany. Due to industrial decline in the area, the 33 metre high water tower lost its function and is now under monument protection and represents an important landmark. It could not easily be reused although part of it was transformed into a space for artistic and creative work (Cultural Heritage in Action, 2021). In other cases, old water towers have been transformed into water museums such as those in St. Petersburg in Russia (St. Petersburg, Museum of Water, n.d.), Kyiv in Ukraine (Kyiv Water Museum, n.d.) and Mulheim an der Ruhr in Germany (Aquarius Water Museum, n.d.).

Linking to the authors' study, it was noted that in the field of water resources the term 'water tower' is sometimes used to describe large water resource catchment areas or glaciers. These towers have an essential role in the Earth's system and are especially important in the global water cycle (Immerzeel *et al.*, 2020, p. 364).

Myth 8: Water is often wasted

Figure 3.7 Some people may use water abundantly.
Illustration: Pertti Väyrynen.

Frequently, it is argued, at least in high-income countries, that water should not be wasted (Figure 3.7) as some people in less developed countries and unplanned areas do not have access even to basic services. As such this is true. although, the idea that saving water in one country would automatically help the situation in another is false. To support low- and middle-income countries, there is an alternative way... as shown in Myth 21.

The term specific water use (SWU) normally refers to daily water use per capita divided by the population (litres per capita per day). In a seminar held in Finland in 1963, it was noted that 'the growth of water use will have no limit', referring to the USA, where SWU was some 500–700 litres per capita per day at that time (Uusi Suomi, 1963). In fact, earlier on, people often talked about water consumption, as if one could consume the water. Thanks to the hydrological cycle to a large extent, it does not vanish, but it rotates – both in the human body and in nature.

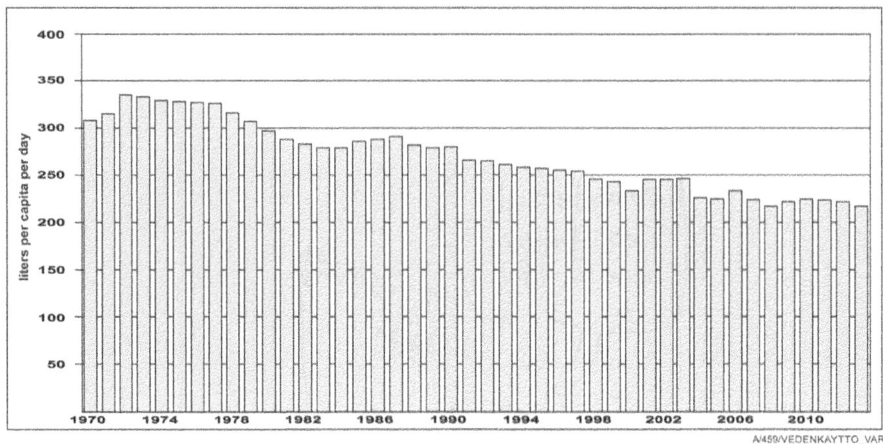

Figure 3.8 Specific water use in Finnish water utilities, 1970–2014.
Source: Lapinlampi (2021, p. 22).

The first energy crisis in Finland was in 1973 with the Wastewater Fee Act later enacted in 1974. This more than doubled the average water price (Myth 11) and had a dramatic impact on declining water use. Thereby, people started to use water more efficiently and water utilities became vastly more interested in controlling water leakages and non-revenue water (NRW). The highest specific water use of 340 l/cap per day in the country occurred in 1972. Thereafter, the rate declined continuously although this has been relatively less in recent years (Figure 3.8). In 2014, this rate was 221 l/cap per day showing a more than 40% decline over these 42 years. After 2014, a slight decrease occurred.

In other countries, such as in Munich, Germany soon after the second world war, the SWU started to decline, most probably due to water leakage control, and again in the 1980s. The declining trend in Finnish and Swedish cities was followed in 1990 by those in Lithuania, Hungary and Poland as a consequence of political change (Figure 3.9). This was partly due to greater efficiency and increased water tariffs, but also to the collapse of many industries connected to public water systems (Juuti & Katko, 2005, pp. 229–230).

REALITY: The efficiency of water use in many countries has improved remarkably since the 1970s and 1980s. In Finland, the energy crises and the 1974 Wastewater Fee Act that more than doubled water prices, created an interest in avoiding water wastage. Since then, the per capita per day use has decreased by an average of 35%. A similar pattern was noted around the same time or slightly later in several other OECD countries. Attention should, however, be paid also to the use of warm water.

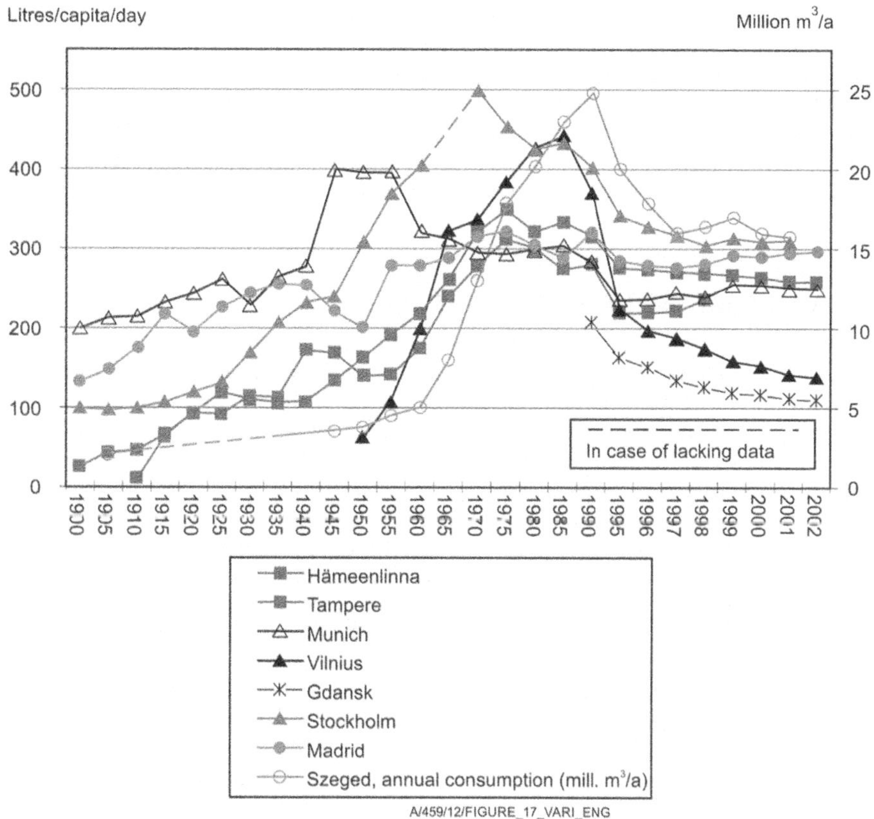

Figure 3.9 Specific water use in selected European cities, 1900–2002.
Source: Juuti and Katko (2005, pp. 229–230).

Regarding water fixtures, the first single lever faucets in Finland came onto the market in the 1950s. Plastic taps covered by chrome were introduced in the 1970s, and the first thermostatic faucets in the 1980s. These were followed by electronic touchless faucets and in the 2010s touchless smart faucets. Development has been continuous, and the next improved version can only be imagined. Along with technological design, user-friendliness and security have improved (Oras, n.d.).

Water saving is still a controversial issue. In cases where a community network has previously been overdesigned, water may not flow fast enough, potentially leading to quality problems due to excess detention. In sewers, smaller amounts of water may not adequately flush away suspended materials. In growing cities, overdesigned pipelines make it possible to expand the networks and use a variety of renovation methods (Myth 10). Along with water saving, more attention should be paid to sewer leakages as the volume of wastewaters to be treated would be less, if leakages were reduced.

It is important to recognise, however, that overuse of water can still have other negative impacts. The more water is used, the more water and wastewater treatment facilities capacity and chemicals are required, and consequently, the more energy for chemical production and transport, for pumping and for producing larger water and wastewater pipes may also be required. In addition, the 'over-abstraction' of the raw water source and the 'surplus' wastewater disposal may cause more harmful impacts, especially on land use and aquatic ecosystems.

3.3 WATER METERING AND WATER USE

Historically, two traditions have prevailed on how to define payments for water use. To the authors' knowledge, in Germany, USA and most other countries, metering of water use has been in operation since the late 1800s. In England, for a long time, payment was based on the ratable value of the property, thus not on actual water use. In most European and North American countries, this metering practice has been in place since the 1880s. In European countries like Norway and more recently in England, metering was introduced, although much later than other countries (Stadtfeld & Schlaweck, 1988). Another European exception is Ireland where metering has been strongly opposed, even until the 2010s (McDonald, 2014, Myth 14).

According to the European Federation of National Associations of Water Services (EurEau, 2021, p. 18), the average water use was 124 l/c/d in the EurEau Member States. Compared to their earlier survey in 2017, average use per capita per day has decreased by 3%, whereas the average annual use per household has decreased by 6%.

In Denmark, the average amount of water a person uses per day in a household has declined from 170 l/c/d in 1976 to 105 l/c/d in 2021 (DANVA, 2022).

From 2005 to 2015, the USA saw a significant decline in surface freshwater withdrawals, accompanied by a 10% drop in household water use, despite an increase in the population of 8% (Hubbart & Ross, 2024). According to Gleick (2023, p. 225), efficient water use is a means to a better future: the focus needs to shift away from using water as efficiently as possible towards benefits that people want – such as good health, food, clean clothes and homes, and other goods.

The Finnish Work Efficiency Institute has explored the trends of specific household water use (SHWU). In 1977, the average SHWU value was approximately 155 l/c/d, most of which was for personal hygiene and toilet flushing. In 1988, the average figure was 116 l/c/d falling to 113 l/c/d in 2020 (Figure 3.10). Thus, from 1977 to 2020 the SHWU value has declined by one fourth (Korhonen *et al.*, 2020, p. 20), which is slightly less than the decline of SWU in communities (Figure 3.8).

Over the period 1977 to 2020 the share of personal hygiene has been almost half of the total use, while use of water in kitchens and toilets has slightly declined relatively. For instance, automated dishwashers have become more

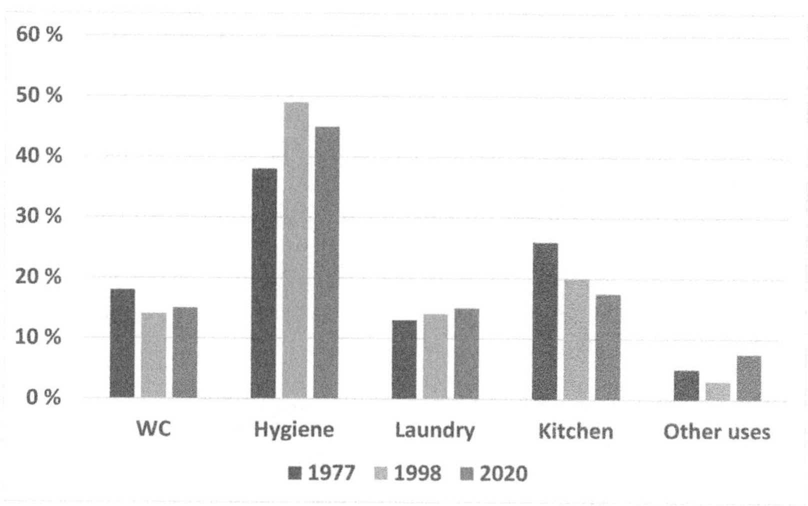

Figure 3.10 Household water use for various purposes in Finland in 1977, 1999 and 2020. *Source:* Korhonen *et al.* (2020, p. 20). Illustration: T. Katko 2024.

popular, which use less water than washing by hand (Bradford, 2017; Korhonen *et al.*, 2020, pp. 19–21).

According to Rajala and Katko (2004), SHWU levels of about 120 l/ capita/day, or even less, can be achieved, while maintaining a high standard of service levels. In addition to individual metering, proper management includes the introduction of modern water fixtures and in-house piping, raising consumer awareness, and active follow-up. The SHWU rate was the lowest in single family houses with water meters and direct consumption based billing. Although decreasing consumption and the resulting fall in water sales may be challenging to water utilities, in the long run it will be feasible to use water wisely even in regions with abundant resources. The more familiar citizens are with the metering and billing system, the more concerned they become about water saving (Figure 3.11).

A study of water use efficiency (FIWA, 2020) pointed out that it is more economic to reduce water leakages than water use. Considering greenhouse gases, it is clearly more feasible to reduce the use of hot water.

It is easy to waste water and it is perhaps likely that young people in particular might use more water than other age groups. In any case, water is seldom wasted on purpose and people should not be blamed for their moderate use of it.

Myth 9: You can throw anything down the toilet

Figure 3.11 Waste that does not belong in toilet bowls, too often tends to be dumped there. *Illustration:* Pertti Väyrynen.

In her classical book *Purity and Danger,* Mary C. Douglas (1921–2007), a cultural anthropologist, notes that waste is substance in the wrong place. Furthermore 'there is no such thing as absolute dirt: it exists in the eye of the beholder'. She identified the concern for purity as a key theme at the heart of every society (Douglas, 1966). Occasionally, powerful world leaders have figuratively warned that they would flush their enemies down the drain. In cinema and books too often there are examples of throwing the most amazing artefacts down the toilet bowl and thereby disseminating misinformation.

In recent years, fat, oil and grease (FOG) blockages, commonly referred to as 'fatbergs', have intensified in sewers. In fact, the term was included in the Oxford English Dictionary in 2015, alongside 'Brexit' (Engelhaupt, 2017). These FOG formations lead to sanitary sewer overflows, property flooding and contamination of water bodies with sewage (Wallace *et al.*, 2017). Fatbergs have formed in sewers worldwide, with the rise of disposable (so-called 'flushable') cloths. Several prominent examples were noted in the 2010s in the UK, with their formation accelerated by ageing Victorian sewers (Fatberg, 2024).

In January 2019, a fatberg of approximately 60 metres in size – blocking the sewer – was discovered in the town of Sidmouth, UK. Schaverien (2019) reported that this was a lump of fat aggregated with wet wipes, sanitary towels and other household products that should be put in the waste bin and not down the toilet. Earlier in 2017, a fatberg was noticed in East London that estimated to weigh 130 tonnes and stretched for 250 metres (Taylor, 2017). Similar phenomena may happen in the level of real estates, though on a smaller scale.

A survey by Kurki (2010) of 38 wastewater treatment plants in Finland revealed the following items that had arrived at the plants via the sewers: soccer balls, mobile phones, children's toys, parts of mattress, glasses, artificial dentures, clothes, and medicine, which presents its own specific problem. At that time it seemed that this problem was declining.

A further survey assessed the labelling, drain line clearance and disintegration testing of 101 consumer products, adopting the International Water Services Flushability Group's specifications for flushability. None of the products tested satisfied the product labelling Code of Practice, and all products other than bathroom tissue failed the disintegration test. Thus, the need for a global definition of a 'flushable' product was found to be necessary (Joksimovic *et al.*, 2020).

For the effective operations and maintenance of wastewater treatment plants, it is important that such items are not conveyed there. Indeed, fatbergs and other disruptions are costly to remove, and public awareness campaigns about flushable waste have been instigated (Fatberg, 2024). One of these included the message to flush only the three P's: Pee, Poop and Toilet Paper (NACWA). Another noted that if solid biowaste is cast into sewers, the rat population living in them will increase.

REALITY: Networks and pumping stations may become clogged due to foreign objects thrown into toilet bowls, resulting in high repair costs. Small objects too, such as cotton buds, strands of hair, and face masks may cause problems. Only urine, excreta and toilet paper are allowed to be put into the toilet bowl.

EurEau; the European Federation of National Associations of Water Services, has clearly stated that we must reduce the release of single use plastics (SUP) in particular into the water cycle. These often end up in the environment and contribute to marine littering (EurEau, 2018). More recently, EurEau (2025) was concerned about the draft international standard ISO/DIS 18671 (i.e. test methodologies for assessing products suitable to be flushed down a toilet and their appropriate labelling), which could have detrimental effects on wastewater systems and the broader water environment by encouraging or enabling inappropriate items to be flushed into sewer networks.

In 2009, a national campaign to prevent inappropriate waste being put into toilet bowls was organised by the Finnish Water Utilities Association; this was remarkably popular. Similar later campaigns have been translated into several languages (Katko, 2016, p. 156; Hörkkö, 2023). For several years, a national Christmas campaign encouraged people not to pour the fat from cooking a ham down the sink but to take it to a collection point (Neste, 2023). This is used to produce biodiesel, with the profits donated to charity. In the global context, the five largest biodiesel producers are Indonesia, USA, Brazil, Germany and China (Insider Monkey, 2023).

In communities, fast-food restaurants produce a lot of fat, and in the USA there are an estimated 160 000 such restaurants serving an estimated 50 million people daily. Grease traps, oil separators and proper grease disposal are important so that they do not become a major pollutant and problem for the sewerage system (Rosinski, 2014). It is estimated that in England and Wales, there are approximately 200 000 sewer blockages and pollution incidents per annum, of which up to 75% are the result of fat, oils and grease (Merton, n.d.).

A common view in Nordic countries is that any excess food wastes should be collected to go to biodegradable waste sites before dishwashing; this is already a common practice and is often available. Another approach is the use of a kitchen sink drain directly to the sewer, which is common in the USA (Cavendish, 2024).

In general, it seems obvious that education and awareness campaigns for stakeholders (citizens and industries) are the foundation for all FOG management initiatives. However, they still need to be repeated often. The Water Museums Global Network (n.d.) linked to UNESCO offers a wider opportunity for raising awareness on water.

Myth 10: The older the water pipes and sewers, the worse their condition

Figure 3.12 Water has been gravitated over long distances to meet urban needs since the roman times. But will the current pipes last as long as the aqueducts?.
Illustration: Pertti Väyrynen.

It is natural to assume that the water pipes and sewers will automatically deteriorate over a certain timeframe in all conditions, after which they need to be replaced. In many countries, the oldest pipes and sewers are over 100 years while in countries like Finland most pipes and sewers were installed after the second world war. However, the lifetime of pipes and sewers is not straightforward due to varying soil conditions, seasonal temperature changes and the construction methods and materials used.

Cast iron pipes constitute the most conventional water pipe material in urban areas. Formerly, these were made from grey cast iron and more recently, from ductile iron. With the exception of plastics, all the materials currently used for pipes to convey water have been in use for centuries. As a historical example, the principal use of stone pipes in Roman times was in aqueducts (Figure 3.12), and particularly for siphons. Lead was used in Roman water pipes and since the Middle Ages a variety of wooden pipes have been used (Campbell, 2021).

The use of lead pipes is a unique case. Several Roman writers had warned about the health risks of lead and recommended using ceramic pipes. However, lead pipes are still used in many countries. After tests in the 1880s, it was found that Finnish surface waters are too acidic and would erode lead pipes. Therefore, it was used only to join cast iron pipes and as pipe material in rare cases inside houses. This is contrary to the situation in Central Europe where groundwaters are rich in calcium, unlike in Fennoscandia (Katko, 2016, pp. 41–43) and, therefore, would not deteriorate.

A more recent case occurred in Michigan State in the USA in the mid-2010s when excess lead was dissolved into surface water that was reused. These events were due to poor decision making, led by non-water professionals (IWA Source, 2016). In the European Union, new legislation is under preparation aiming to reduce lead in valves and fixtures from the current maximum standard of 10 μg/L at the tap to 5 μg/L (Wavin, 2024).

In more modern times, in countries such as Finland, plastic pipe manufacturing began in the 1950s, initially of Polyvinyl Chloride (PVC) pipes, and later Polyethylene High Density (PEH) or Acrylonitrile Butadiene Styrene (ABS) pipes. Plastic pipes soon became the most dominant material both in water supply and sewerage systems in Finland. Polypropylene pipes (PP) have also been used in drinking water and cross-linked polyethylene such as PEX, XPE or XLPE, in building services.

At first these pipes were used for rural pipelines, and thereafter in larger communities, and later in major cities. Currently, Finland probably uses more plastic pipes in relative terms in both water and sewage systems than any other country (Katko, 2016, p. 45). In urban water networks, cast iron and later ductile iron pipes have been the most traditional pipe material (Katko, 2016, p. 88). Typically, for larger pipes and in areas of high traffic load ductile iron is used. In sewers, concrete pipes used to be dominant especially if the city owned a concrete foundry. Gradually, PEH pipes have become more popular in sewers apart from for the larger size pipes (Katko, 2016, p. 90).

The use of plastic pipes for domestic water has traditionally been treated with more caution in the USA. While industry experts confirm that plastic is a safe material with which to replace lead pipes, some researchers and health advocates are not so sure (Lloyd, 2022). In California in 2007, a twenty-year battle ended in favour of housebuilders, who won the right to install chlorinated polyvinyl chloride plastic water pipes rather than copper (The Mercury News, 2007). According to the European Plastic Pipes and Fittings Association, the lifetime of the plastic pipes can be more than a hundred years assuming that all stages of design, manufacturing, trenching, and operation follow valid standards (Pinter *et al.*, 2024).

REALITY: The age of network material does not necessarily say much about the actual condition of the network. Pipe materials used, installation methods, and soil conditions can shorten their service life significantly. However, networks do not last hundreds of years and, therefore, their rate of rehabilitation needs to be increased.

In Finland, approximately 90% of water supply networks have plastic pipes, the rest being mainly made of ductile iron. This has grown significantly since 1970, when the figure was only 45%. In sewers, the proportion of plastic pipes has increased and that of concrete has decreased. Accordingly, the percentage of concrete sewers declined from almost 90% to just over 10% over the same period (Lapinlampi, 2021).

A major advantage of polyethylene (PE) pipes is that they can be joined by welding. Both water pipes and sewers can be installed to go under water bodies making it easier to expand networks. This has made it possible to collect sewage from large areas and centralise its treatment, which has both its pros and cons.

As discussed in Myth 4, conventionally there are two types of systems for collecting wastewaters – *separate and combined*. In a separate system wastewaters are collected to wastewater treatment plants while stormwaters are normally discharged to the nearest water bodies without treatment or possibly with physical treatment only (Botturi *et al.*, 2021). It is likely that with increasing weather fluctuations there will be additional needs to treat at least the first flushes of stormwaters. In many older cities and their central areas, sewers are combined while newer areas have separate systems.

In Europe, the relative share of combined sewers varies quite a lot, for instance, in Austria it is one fourth, in Germany close to one half, in Italy they are extensive, while in Nordic countries they make up only a minor part of the total sewers (EurEau, 2021, p. 23).

In countries with winter conditions and frost, soil water, wastewater and stormwater are normally located within the same excavation, often approximately 2.5 metres above the ground surface. In a vertical direction, water pipes are ideally at the top, stormwater pipes somewhat lower and wastewater sewers are lowest. In developing economies, it is common to locate water pipes and sewers on the same open concrete channel as the stormwaters. This is practical but contains risks of contamination.

Asbestos cement pipes were popular drinking water distribution systems for example in the mid-1900s in the USA and the 1970s in Finland. It was used because of its light weight, low coefficient of friction and resistance to corrosion. Unfortunately, it was later found that people who use water with high amounts of asbestos over extended periods may face increased health risks such as cancer (Stenstedt, 2019). Its production was stopped in Finland due to the dangerous dust created in its manufacturing and dismantling (Juuti & Katko, 1998). However, in some places asbestos cement pipes are still used as their complete removal will take time.

3.4 AGEING INFRASTRUCTURE

Ageing infrastructure and the need for additional reinvestment in water and wastewater infrastructure presents a major challenge in many countries. In Finland, at least two or three times the current rate is needed (Hukka & Katko, 2015a, 2015b). Based on national information and benchmarking systems, it was estimated that the total investment needed was nearly double for the coming 20 years. Of this approximately 60% was required for network rehabilitation. The share of completely new investments was estimated to be approximately 26% (Kuulas *et al.*, 2020, p. 76).

According to Najar and Persson (2023), in Sweden, only 4% of 184 organisations explored met the requirements for a sustainable water and wastewater facility according to the sustainability index (SI), which is an element of the Swedish benchmarking system.

The asset renewal rate for wastewater infrastructure in Europe varies a lot from one country to another according to the local water management system, the age of the infrastructure, the impact of depreciation and the level of requirements (Eureau, 2021, p. 28). For instance, in Finland it is estimated that the renewal rate was close to 0.7% per year (Eureau, 2021, p. 28).

In OECD countries, substantial investments over many decades have largely ensured access to safe water supply and sanitation. Significant investments will, however, be necessary to rehabilitate the existing infrastructures, so they conform to more stringent environmental and health regulations, and to maintain service quality in the future. In non-OECD countries, the challenges are more daunting, since a large share of their populations have no access to services, and many suffer due to their unsatisfactory quality.

Table 3.1 presents a summary of the state of water service infrastructure and estimated funding gaps in Canada, Finland, Norway and the USA for 2013 and 2014. The funding gaps do not, however, cover the rehabilitation, renewal and replacement needs of stormwater infrastructure and house connections. The latter is commonly assumed to be one third of the total. Although the data are not directly comparable, they indicate the severity of the problem.

The coverage of improved water services in sub-Saharan Africa is still rather low in spite of progress especially relating to water supply from 2000 to 2015 (Armah *et al.*, 2018). In 2020, access to basic drinking water was 69% and to safely managed water was 39%. The rate of no open defecation was 89%, whereas safely managed sanitation was only 27%. Thus, to achieve SDG WASH targets in Africa would require a dramatic acceleration in current rates (WHO *et al.*, 2022). Due to common operational and efficiency problems, the actual service coverage figures are lower (Behailu *et al.*, 2016; Biswas, 2022).

In Canada in 2019, approximately 30% of water infrastructure such as watermains and sewers were in fair, poor or very poor condition (The

Table 3.1 Estimated investment gap in water services infrastructure in selected OECD countries. (Hukka and Katko, 2015a, 2015b, modified).

Country	Population	State of Water Services Infrastructure (Scale used)	Estimated Funding Cap Per Capita (USD)
Canada	35 182 000	Good (Very good–very poor)	2 050
Finland	5 426 000	7 (10–4)	1 210
Norway	5 043 000	Water networks 3 (5–1) Sewers 2 (5–1)	3 590
USA	320 051 000	D (A–E)	1 740

Note: Currency exchange rates on 3 March 2014; Population figures: UN-DESA, 2013.

Association of Consulting Engineering Companies Canada *et al.*, 2019). The data from 2017 and 2019 on European water utilities show that the renewal rate of wastewater infrastructure varies from 0.1 to 1.5 (%/year) (Eureau, 2021, p. 28) the mean value being 1.6. In the USA, the state of drinking water from 1998 to 2017 as measured every three to four years, has been D or D−, while that of sewage is D or D+ (Michelson, 2023).

Using digital algorithms for optimisation of water distribution systems is becoming increasingly relevant. Through real-time optimisation it is possible to gain remarkable savings in the use of energy and chemicals. This method seeks most feasible flow and pressure settings to all water production units as well as pump and valve stations. It considers actual and forecasted water use and water production rates, pumping capacities and limitations, and pressure limits, and strives to use water reservoirs as efficiently as possible (Sunela, 2024; Sunela & Puust, 2017). Such developments are also reported by Batista do Egito *et al.* (2023) and Awe *et al.* (2019).

The lifetime of various pipe materials may vary a lot. With good soil conditions, materials and proper installation, they may last over a century while in conditions that freeze and melt, their life may be much shorter. Due to the variety of materials and methods used, renovation and rehabilitation of networks can be quite challenging.

All things considered, water pipes contain a vulnerable benefit, the nature of which differs fundamentally from electricity, heat or telecommunication networks. Water pipes need to deliver its invaluable product from one place to another safely, ensuring water quality and allowing the used water to be treated efficiently before discharging the final effluent to water bodies.

Myth 11: Wastewater treatment plants are major polluters

Figure 3.13 Sewer divers are sometimes needed but probably not like this! *Illustration:* Pertti Väyrynen.

As mentioned earlier, the available sewerage system, either combined or separate, has a strong impact on whether wastewaters can be efficiently treated or not. Occasionally, it is argued that wastewater treatment plants are major polluters of water bodies. In the early phase of their development, this might have been true, or could still be the case where treatment does not yet exist. If wastewater treatment is advanced, it will remove organic loading (normally measured as biological oxygen demand – BOD) and major nutrients, nitrogen (N) and/or phosphorus (P) very efficiently. Even at a high removal rate, it is particularly important to find the most appropriate site for the final discharge to water bodies. If possible, there should be minimal effect of loadings on the downstream water body, which is often not that easy. In some cases, divers are needed to maintain the sewers, although not in the way depicted in Figure 3.13.

In lakes, phosphorus (P) is the *minimum nutrient* for eutrophication (Eutrophication, 2024) of water bodies; this is the gradual increase in the concentration of phosphorus, nitrogen, and other plant nutrients in an aquatic ecosystem. However, the removal of nitrogen seems to be a controversial issue. The European Union via its Water Framework Directive (European Commission, n.d., b) has required both phosphorus and nitrogen removal relating to the Baltic Sea and other coastal waters. In the Finnish case, there is confusion over nitrification. Similar recommendations have been made for eastern coastal areas in the USA (Grigg, 2024).

Despite recent claims that reducing nitrogen is essential to curb estuarine eutrophication, Schindler (2012) in Canada argued that there is no documented case history of where this has been successful. In addition, Schindler *et al.* (2008) stated that experiments to guide nutrient management confidently must be on a full ecosystem scale and carried out for a period of several years. In certain conditions, nitrogen removal may even be harmful in terms of eutrophication. In any case, these experiments show how the conditions may vary in nature and, therefore, case by case consideration is useful.

According to marine experts Tamminen and Andersen (2007), along the coastal zones of the Baltic Sea, efforts to limit nutrients may be restricted as they vary depending on the situation and the season. In Finland, largely based on these studies, nitrogen removal (as NH4) has been required at wastewater treatment plants serving over 10 000 people except in the northern part of the Gulf of Bothnia (Knuuttila, 2024). Currently, this is also a requirement in several inland cities.

Lappalainen (2018), as a limnologist, criticised the prevailing view that anthropogenic (i.e. caused by humans or their activities) nutrient loading is almost the sole cause of eutrophication. He pointed out the role of internal loading in the Baltic Sea, which phenomenon has been noted also by some authorities (Marine Finland).

According to the Baltic Sea Centre (2021), the state of the Baltic Sea would be significantly worse if nothing had been done to reduce nutrient loadings. Had these stayed at the level of the 1980s, phosphorus content would be 40% higher and amounts of algae approximately 50% higher than at present.

In Sweden, large plants have typical treatment requirements as <0.3 mg/l P, <10 mg/l N and BOD7 <10 mg/l. Some new environmental permits have been issued for <8 mg/l N. According to the environmental code, the regulation should be set for each plant individually. In practice, the parametric values are the same for plants with discharges in the same order of magnitude (Persson, 2024).

In the USA, according to EPA (2024), progress on wastewater treatment is viewed according to five categories as follows: level five with a complete set of N and P for all water types (in five states); level four with two or more water types with N and/or criteria (in three states); level two with some waters with N and/or criteria (in 15 states); and level one with no N/P criteria (in 25 states).

> **REALITY:** Modern wastewater treatment plants are often among the largest environmental investments within communities. Biological oxygen demand (BOD) and for example, in the case of Finland, the major nutrient, phosphorus, are reduced by over 96%. Regarding treatment requirements, the needs of the recipient water body should be considered. However, experts seem to have different views on the necessity of nitrogen removal.

The question of minimum nutrient(s) in water bodies in various conditions is problematic in itself, since even the experts have contradictory views on the matter. Once more research has been conducted and experience gained, it is possible that unexpected findings will later emerge.

The aim of the Urban Wastewater Directive of the European Union is to protect the environment from being adversely affected by insufficiently treated urban wastewater discharges. At the time of writing this, the directive is under revision, in accordance with the One Health approach aimed at balancing and optimising the health of people, animals and ecosystems, with treatment requirements to be increased (Council of the European Union, 2024).

As an example of successful wastewater treatment and water pollution control policy, the authors present the development of wastewater treatment in Finland and its major legislation since the late 1800s to the early 2000s (Figure 3.14). The initial weak signals were the Health Decree of 1879 and the Water Rights Act of 1902. The former involved the first mandated survey of city elevations as the basis for sewerage planning, while the latter focused on economic utilisation of water bodies (Katko, 2016, p. 45).

The major driving forces for pollution control included the acceptance of flush toilets around 1900, the Health Decree of 1928, the construction of the first separate sewers in 1938, and EU membership in 1995. The key decisions, including binding requirements for wastewater treatment, were based on three specific acts: the Water Act 1962, the Wastewater Fee Act 1974, and the Water Services Act 2001.

The Water Act enacted in 1962 stipulated that municipalities had to apply for permits to discharge their wastewaters, which included requirements for wastewater treatment. The Wastewater Fee Act in 1974 meant that in most cases water fees more than *doubled*, which to the authors' knowledge is a unique case. This raised very little opposition since everybody could see how municipal and forest industry wastewaters had polluted the nearby water bodies (Katko, 2016, pp. 48, 69, 105, 136, 139). This is quite different from the policy applied, for example in Sweden and the USA, where the first wastewater treatment plants were mainly financed by public grants (Juuti *et al.*, 2009). Between 1973 and 1981, federal grants covered up to 75% sewage of the treatment plant construction costs in the USA (Sedlak, 2014). In many other countries, these developments were largely financed through taxation before introducing possible wastewater fees.

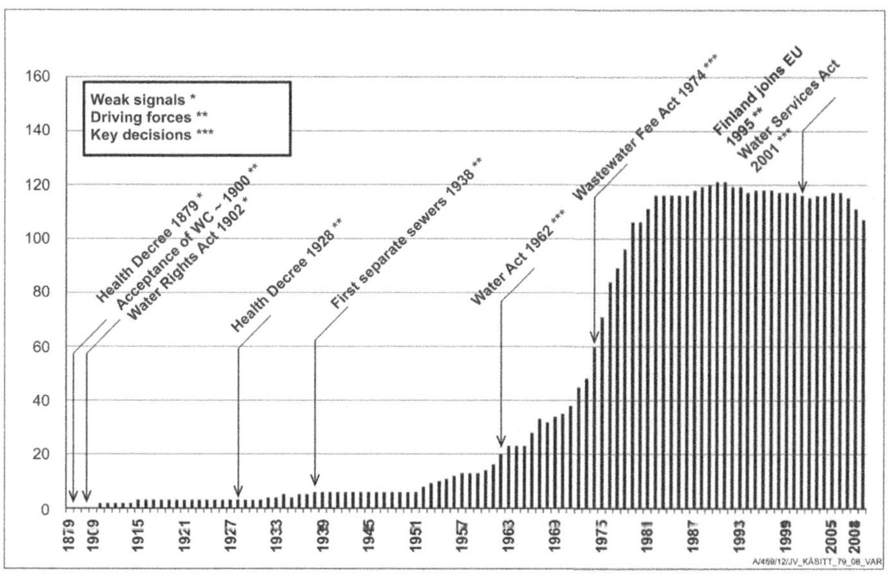

Figure 3.14 Total number of urban wastewater treatment plants in 1910–2009 in Finland, the number located in the 102 cities in existence in 1994, and related key legislation. *Source:* Katko (2016) p. 105, illustration derived from several sources.

The actual boom in wastewater treatment plant (WWTP) construction in Finland happened from the mid-1960s to 1980s thanks to proper legislation and control. Since 2000, the number of WWTPs has decreased due to the expansion of sewerage systems and construction of advanced and more centralised WWTPs. Since 2021, several Finnish WWTPs have signed voluntary Green Deals with the Ministry of the Environment. Under these operational commitments, the WWTPs have promised to attain even better results in water protection than those required by authorities (Anon, 2021).

In many countries, water pollution control expanded in the 1960s and 1970s. Although several of the current European Union member countries moved well ahead, others such as Milan, one of the richest European industrial centres did not start treating its wastewaters until autumn 2004 (Juuti & Katko, 2005, p. 226) In Brussels, the headquarters of the European Union, the first wastewater treatment plant for the southern part of the city came into service in 2000 (Katko, 2016, p. 109), although the city still faces challenges with water pollution control (Zimmermann, 2022).

In addition to community wastewaters, there are several other loading sources and types. Figure 3.15 shows the development of phosphorus loadings from point and non-point sources in Finland from 1991 to 2019. In principle, it is easier to get control of point sources while non-point loadings come from dispersed sources such as forests, marshland and agriculture.

(a)   P loadings from point sources to water bodies in 1991, 2008 and 2019 (t/a)

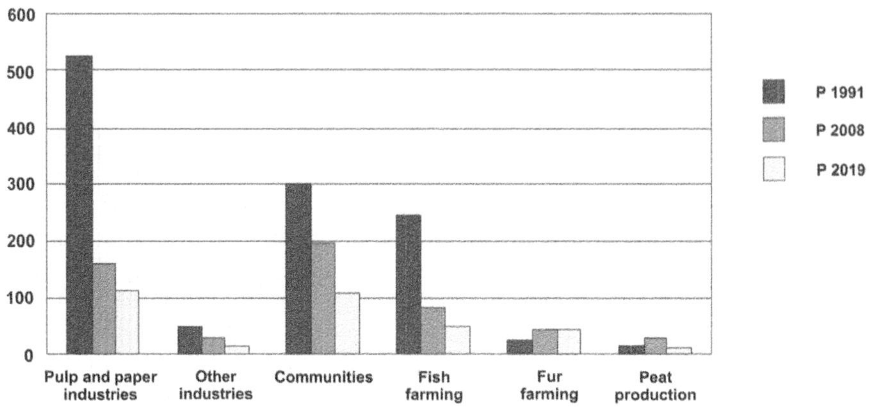

(b)   P loadings from non-point sources to water bodies in 1991, 2008 and 2019 (t/a)

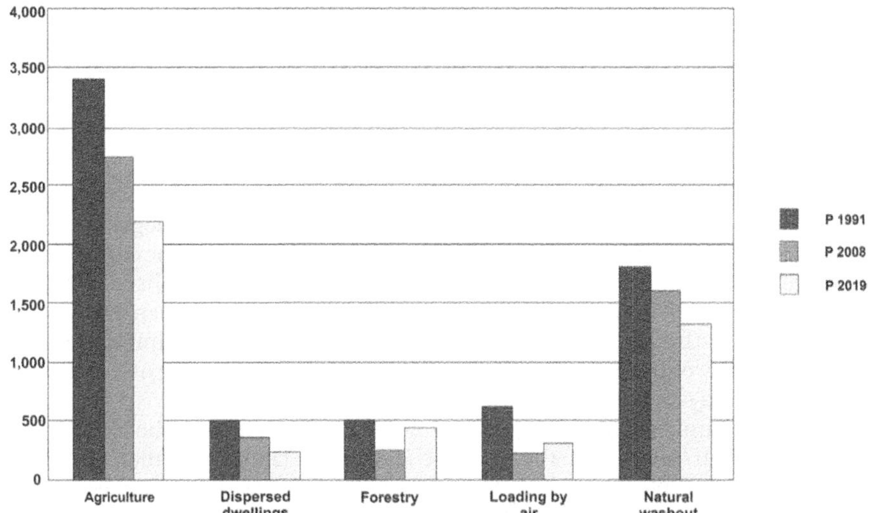

Figure 3.15 Phosphorus loadings to water bodies from (a) point sources and (b) non-point sources in Finland in 1991, 2008 and 2019.
Source: SYKE (2021).

As shown, loadings from point sources have declined remarkably by almost a third in some communities and by even more in others. From non-point sources, loadings have also declined but to a relatively lesser amount. It is good to remember that we also have a notable share of natural washout.

In 2006, Finland and Sweden were summoned to the European Court of Justice (ECJ) for the failure to systematically remove nitrogen from wastewaters of inland cities and towns, thus contributing to the environmental problems of the Baltic Sea. The Finnish authorities' view, however, was that nitrogen removal was adequate considering the recipient water body (Katko, 2016, pp. 108–109). The ECJ ultimately dismissed the EU Commission's action against Finland and ordered the Commission to pay Finland's court costs (YLE, 2009). The court's view was that the Commission was unable to show that full nitrogen removal was required for all treatment plants serving a population equivalent of greater than 10 000. At the current removal rate of nutrients (P and N), the impacts of eutrophication are clearly lower than if the rates of the directive (N 70% and P 80%) were applied (Kaloinen, 2015). Longterm studies by Finnish and Swedish water protection associations have shown that the nitrogen of inland water bodies is largely absorbed before reaching the sea (Katko, 2016, pp. 108–109).

Similar to pulp and paper manufacturing industries, there has been a lot of progress in water pollution control in Finland. From 1950 to 2007, the production of paper and cardboard had grown by about twenty-fold, and the production of pulp by almost ten-fold. By the early 2020s, the production of paper had declined to one third, whereas that of chemical pulp and cardboard had grown by a quarter (Finnish Forest Industries, 2024).

The loadings from forest industries began to decline in the 1970s along with water pollution control measures. At that time, a sulphite pulp mill could use approximately 350 m³ of water per produced ton, and the older sulphite method even more (Kytö, 1974), the lowest current figures are approximately 5 m³ per produced ton thanks to better processes and internal cycles, thus showing a very strong improvement in the efficiency of water use.

On the European scale, approximately 40% of surface water bodies have good ecological status and 38% of them have good chemical status. In addition, approximately 85% of bathing water sites are of excellent quality and over 90% of urban wastewater is dealt with in line with EU standards (European Commission, 2024).

A more recent challenge related to water pollution comes from mining. In South Africa, Grewar (2019) reviewed options for reusing mine impacted waters. There, the government believes that substantial volumes of water would be available through the reuse of treated mining-impacted waters. Currently, large volumes of such waters are produced, often with no end use. Treating these wastewaters to meet 'fitness-for-use' guidelines would reduce the treatment costs compared to potable water production. The review suggests that the most suitable option for reuse of treated mine impacted water is for irrigation of a variety of crops.

However, globally, the statistics for wastewater treatment are stark. Corcoran *et al.* (2010) referred to the term 'sick water' when noting that around 2010, two million tons of sewage, industrial and agricultural waste was discharged into the world's waterways and at least 1.8 million children under five years old died every year from water related diseases; this is one every twenty seconds.

According to the United Nations World Water Development Report 2017 (WWAP, 2017), over 80% of the world's wastewater was released into the environment without any treatment. High-income countries treated approximately 70% of their municipal and industrial wastewaters. In upper-middle-income countries this ratio dropped down to 38%, and in lower-middle-income countries to 28%. In low-income countries, only 8% had treatment of any kind.

In 2020, it was estimated that 56% of global household wastewater flows was safely treated (UN-Habitat and WHO, 2021). For regions, Jones *et al.* (2021) reported that wastewater collection and treatment rates are the highest in western Europe (88% and 86%) and the lowest in South Asia (31% and 16%) and sub-Saharan Africa (23% and 16%). The share of untreated wastewater is still far too high in many parts of the world. Yet, it could be that due to operations and maintenance problems, the actual levels of wastewater treatment might be lower than the figures presented.

Waste stabilisation ponds (WSPs) are man-made earthen basins used for wastewater treatment that aim to reduce nutrients, micro- and macro-pollutants and remove pathogens from wastewater. According to Kadri *et al.* (2021), WSPs would be especially suitable for low-income countries that have warm climates as they are cost-effective, highly efficient, entirely natural and highly sustainable.

At a global scale, there is a long way to go in water pollution control. It is clear that WWTPs are not water polluters but on the contrary, they have done a good job in saving surface waters and the environment. Public relations in the media are important in promoting awareness of these issues.

It is likely that step-by-step development is the most feasible way forward in most cases. As an alternative to waterborne sanitation, we discuss the option of dry sanitation as part of Property Managed Water Supply and Sanitation in Myth 12.

Myth 12: Only centralised piped water and waterborne sewerage are good enough in a modern society

Figure 3.16 Some of us may dream of making piped water and sewerage available everywhere. But is this really sensible?
Illustration: Pertti Väyrynen.

While myths one to eleven have mainly focused on network-based water supply and wastewater services, this myth is about self-supply systems that are owned and managed either by households themselves or by a small group of households.

Many countries including high-income economies may have substantial rural populations that have no access to public water and/or wastewater systems. This category of Property Managed Water Supply and Sanitation (PMWSS) constitutes an additional 'fourth' level that should be taken into account. But how should PMWSS be organised and managed?

In Finland, Sweden, and Norway, in total approximately three million inhabitants live outside the centralised sewer network which represents 13% of their combined population. In addition, approximately half a million second homes exist in each country, the majority of which are situated outside centrally sewered areas (Laukka *et al.*, 2022). Despite the anticipated increase in urbanisation, in the European Union there are large rural areas which make up half of Europe and represent around 20% of the population (European Commission, n.d., a).

It is often thought that a dry toilet is an old-fashioned solution, and a modern apartment should automatically have a water toilet (Figure 3.16).

Today, solutions have been developed that make the use of dry toilets easier and more convenient than ever before. In addition, faeces and urine contain a lot of nutrients and, therefore, can be used as fertilisers. This has obvious advantages, at least for holiday homes, and when fertilisers are in short supply or are expensive.

In many countries waterborne sanitation (the water closet or WC) revolutionised urban and rural lifestyles in the 1950s and 1960s. Some people may have wondered why they had to go out to eat and defecate in the house, since they were used to doing the opposite. However, WCs caused many environmental and hygiene problems at that time in urban and rural areas because there were not necessarily proper systems for treating toilet waters (Juuti & Wallenius, 2005).

In the USA, Maxcy-Brown *et al.* (2023) estimated that approximately 25% of American households use onsite wastewater treatment systems (OWTS); this figure exceeds the American Housing Survey estimate (15.7%) by over 12 million households. They noted as well that the Joint Monitoring Program for Water Supply, Sanitation and Hygiene (JMP by UNICEF and WHO) data on achieving the SDG target 6.2 may underestimate the prevalence of unsafely managed sanitation in high-income countries. Their concern was the potential under-prioritisation of data on a key human right that is not only essential for protecting human health but also for the environment. Furthermore, most nationwide surveys overlook homeless populations and illegal immigrants (authors' note).

As for sanitation infrastructure, latrines are too often viewed with disdain and in many cultures talking about bodily functions is still a taboo. The caretakers of latrines are often seen as the lowest class of all working people. In poor living conditions, the social status of tenants was indicated by their location – the further one lived from the latrine, the better. Nevertheless, it is vital to talk with people about what kind of solutions are acceptable to them (Juuti & Wallenius, 2005).

In 2004, the first actual wastewater decree for sparsely populated areas came into force in Finland, which called for more efficient property level wastewater systems within a ten-year period. Implementation of this decree, however, proved more difficult than anticipated. Nonetheless, its stringency was eased by implementing it in several phases. In 2017, the requirements to renew the wastewater systems of permanent housing became compulsory in areas close to water bodies or situated in groundwater areas. For permanent housing, approximately 16% and 7% of the two categories are situated in such areas; these figures are surprisingly low. Thus, it is probable that implementing the current requirements would have been more viable during the initial phase. Related to these, Kallio (2020) analysed the efforts of on-site wastewater treatment guidance in Finland. While found profound and professional, it remains to be seen whether the improvement of wastewater treatment will need more effective monitoring.

The weakest point of the decree is that the investment, use and maintenance of water supply and wastewater treatment equipment is left entirely to the property owner/occupant. If a water toilet is used, the wastewater should be properly treated. It is not reasonable to assume that the average citizen would automatically be the operator of a wastewater treatment plant. Based on the interviews and discussions that took place, it seems that the wastewater decree in sparsely populated areas was taken forward largely according to the terms of the legislation. In addition to this, it would have been possible to experiment further to identify approaches that work (Mattila, 2005).

REALITY: Property Managed Water Supply is one of the options in reaching SDG6 to ensure access to water and sanitation for all. For sanitation, it is recommended to use dry toilets in areas without water and wastewater infrastructure. Yet, for urban areas and larger scale solutions, such systems are not likely to be feasible.

Based on Nordic and other European countries for governing on-site sanitation, Laukka *et al.* (2022) recommended the following:

(i) A coherent national regulatory framework is of critical importance as the basis for good governance practice and a functioning sanitation service chain.
(ii) Implementation of the regulations requires adequate support mechanisms at all levels of implementation (national, regional and local).
(iii) Public awareness needs to be increased to strengthen the policy relevance of on-site sanitation.
(iv) Common platforms are required for presenting the results of various studies, sharing good practices, coworking and learning from each other both at national and international level.
(v) In addition to those above, regional and local level frameworks, and payment and/or subsidies should likely be considered.

For some years, it has been suggested that alternatives for centralised community water and wastewater systems should be developed (Sustainable Sanitation Alliance). Among others, several of the authors were involved in organising a series of six Dry Toilet Conferences from 2003 to 2018 in Finland (Katko, 2016, pp. 114–115). Accordingly, the IWA hosted its first Non-Sewered Sanitation Conference in 2023 in Johannesburg, South Africa (IWA, 2023). However, particularly in relation to sanitation, historical experiences and cultural heritage need to be taken into account.

In their study on 'self-supply' focusing on sub-Saharan Africa, Sutton and Butterworth (2021, p. 265) stated that in low- and many middle-income countries, self-supply can fill gaps in public water provision, particularly among low-density rural populations. The role of self-supply would need greater recognition and a policy change by governments, development partners and practitioners. Self-supply has long been overlooked since it is not mapped, monitored or regulated. Thereby it is invisible to public policymakers and decision takers.

Sutton and Butterworth (2021, p. 27) further pointed out how self-supply has been part of the historical development in countries that have reached universal access. In this field, the authors have also made many contributions (Cadwes – Books, 2024), as discussed more in chapter five, Myth 18.

The fact that major water service gaps still exist in rural households forcing them to invest directly in self-supply was highlighted by Wainaina and Barbosa (2024). They argued that self-supply is the foundational and often preferred model for rural households due to its reliability, predictability, control and flexibility. Therefore, they called for investment in research on self-supply policy and practice.

In rural Ethiopia, water is still collected in many places from natural or improved spring sources, where water flows freely. A house-specific unprotected dug well is used in areas where the groundwater level is high (close to the surface). Community managed wells are normally equipped with a hand pump or motorised pump. The number of users of protected spring sources and wells ranges from 150 to 500. In piped systems, water flows by means of gravity and is normally distributed through public water points along the canal or pipe. People fetch their water with jerrycans and pay based on the quantity taken. In areas with high fluoride content, water is sometimes treated with bone charcoal filtration. In the countryside there are no water toilets and it is still very common to use a simple pit ('long-drop toilet') with a concrete or wooden cover (Suominen, 2024).

According to Suominen (2024), self-supply has been on the agenda of Ethiopian WASH since 2008 but from 2017 professionalisation and piped water supply started to dominate the water strategy and self-supply was largely forgotten. In 2024, self-supply was again resurrected and approval was given for it to be developed further in Ethiopia.

In their study, Duque *et al.* (2024) stated that decentralised systems can offer economic advantages when capital expenditure costs for small-scale wastewater treatment plants are significantly reduced compared to current costs. While this is possible, as shown in Myth 5, water and wastewater treatment are significantly cheaper than the costs of water and sewer networks. But it is perhaps good to keep in mind their question: 'When does infrastructure hybridisation outperform centralised infrastructure paradigms?'.

The Rural Water Supply Network (RWSN) Executive Steering Committee (2010) presented seven myths on rural water supplies. While the first suggests using public funds and heavily subsidising hardware, the RWSN argued for greater recognition of the financial contributions by households and communities as well as improved knowledge and skills. Second, building rural water supply systems is clearly not more imperative than keeping them working, but both factors are important. Third, since communities are not always capable of managing their own facilities, the following are needed: (i) meaningful participation and training of communities; (ii) capacity, financial resources and monitoring systems to support them, and (iii) testing of alternatives in community management.

In Myth 4, rather than stipulating 20 litres per person per day of clean water, there is an urgent need to consider (i) other water requirements, such as for livestock and crops; (ii) household values with respect to water and (iii) presentation and demonstration of real and affordable choices for household water supplies. Contrary to the myth 'we know what we want and what we can get from the private sector', there are very few incentives for the private sector to invest in construction or management of RWS. Although any action to improve rural water supplies would be laudable, the RWSN paper advocated for: (i) coordination between RWS actors at all levels; (ii) strengthened institutions and improved mechanisms to NGOs, other government agencies and donors; (iii) raising awareness of the damage that can occur with misdirected approaches; (iv) scheduling project implementation for the benefit of rural dwellers; and (v) high levels of professionalism and work ethic. The final point argued in the paper was that there is no quick fix to substitute for long-term political negotiation, institution building, education, investment and innovation. The committee concluded that much more flexible, adaptive approaches are needed (RWSN Executive Steering Committee, 2010).

Mattila (2024) suggests that in property managed water systems and sanitation, active product development would be needed. In areas with occasional droughts, waterborne sanitation should use water as little as possible.

Myth 12 on Property Managed Water Supply and Sanitation (PMWSS) is a further example of how in sustainable water services we cannot merely focus on technology; multifaceted issues and miscellaneous requirements should be taken into account (Myth 15).

PMWSS is one option for water services delivery and thus is a part of the institutional diversity that exists in water services provision and production. This leads us to chapter four, focussing on the myths related to institutional factors and arrangements – the rules of the game. For water and sanitation as a human right, we refer to Myth 14.

3.5 DISCUSSION QUESTIONS

(1)　Evaluate how many times you use tap water, either directly or indirectly, in a single 24-hour period.

(2)　Evaluate how many times you use the toilet in one 24-hour period. By what processes do you think toilets are placed in public spaces?

(3)　What are the advantages and disadvantages of bottled water?

(4)　What kind of water tower designs blend into the landscape and which do not?

(5)　Can the specific use of water still decline in the future?

(6)　What should be done to prevent items that do not belong there from ending up in the sewer?

(7)　How can the renovation and renewal of water and wastewater networks be promoted?

(8)　To what extent is water pollution still caused by human activities?

(9)　What kind of benefits can self-supply water and dry toilets bring to residents?

REFERENCES

ACIR (1987). The organization of local public economies, Washington, DC. Advisory Commission on Intergovernmental Relations; cited by Heikkila T. (2004, 102).

Anon (2021). Sitoumus 2050, https://sitoumus2050.fi/en_US/jatevesi (accessed 22 March 2025)

Aquarius Water Museum (n.d.). https://www.erih.net/i-want-to-go-there/site/aquarius-water-museum (accessed 22 March 2025).

Armah F. A., Ekumah B., Yawson D. O., Odoi J. O., Afitiri A.-R. and Nyieku F. E. (2018). Access to improved water and sanitation in sub-Saharan Africa in a quarter century. *Heliyon*, **4**(11), e00931, https://doi.org/10.1016/j.heliyon.2018.e00931

Asola I. (2001). Vesitorni – yhdyskunnan maamerkki (Water tower–Landmark of the Community) Association of Finnish Civil Engineers, Helsinki. [in Finnish, Captions and synopsis in each chapter in English].

Awe O. M., Okolie S. T. A. and Fayomi O. S. I. (2019). Optimization of water distribution systems: a review. *Journal of Physics: Conference Series*, **1378**, 022068, https://doi.org/10.1088/1742-6596/1378/2/022068

Baltic Sea Centre (2021). Stockholm University. Further land-based measures are needed to reach eutrophication targets. Policy Brief. https://www.su.se/polopoly_fs/1.580504.1635505389/!menu/standard/file/PBinternbelastningEngWebb.pdf (accessed 22 March 2025).

Barnett C. C., Minis H. P. and VanSant J. (1997). Democratic Decentralization. Research Triangle Institute, North Carolina, USA.

Bärthel H. (1997). Wasser für Berlin (Water for Berlin). Herausgeber Berliner Wasser Betriebe. Verlag für Bauwesen Berlin, Germany, p. 304.

Batista do Egito T., Gonçalves de Azevedo J. R. and Bezerra S. T. M. (2023). Optimization of the operation of water distribution systems with emphasis on the joint optimization of pumps and reservoirs. *Water Supply*, **23**(3), 1094–1105, https://doi.org/10.2166/ws.2023.065

Behailu B. M., Hukka J. J. and Katko T. S. (2016). Service failures of rural water supply systems in Ethiopia and their policy implications. *Public Works Management & Policy*, **22**(2), 179–196, https://doi.org/10.1177/1087724X166561

Biswas A. K. (2022). Urban water security for developing countries. *River*, **1**(1), 15–24, https://doi.org/10.1002/rvr2.11

Botturi A., Ozbayram E. G., Tondera K., Gilbert N. I., Rouault P., Caradot N., Gutierrez O., Daneshgar S., Frison N., Akyol C., Foglia A., Eusebi A. L. and Fatone F. (2021). Combined sewer overflows: a critical review on best practice and innovative solutions to mitigate impacts on environment and human health. *Critical Reviews in Environmental Science and Technology*, **51**(15), 1585–1618, https://doi.org/10.1080/10643389.2020.1757957

Bradford A. (2017). Dishwasher vs. hand-washing: What saves more water? *CNET*, 7 March. https://www.cnet.com/home/kitchen-and-household/how-much-water-do-dishwashers-use/ (accessed 22 March 2025).

Burbridge D. (2017). The inherently political nature of subsidiarity. *The American Journal of Jurisprudence*, **62**(2), 143–164, pp. 161–162, https://doi.org/10.1093/ajj/aux017

Cadwes – Books (2024) http://www.cadwes.com/publications/books/ (accessed 22 March 2025).

Campbell J. W. P. (2021). The Development of Water Pipes: a Brief History from Ancient Times until 1800. The Eighth Annual Conference of the Construction History Society. Conference Paper, Sept 2021. https://www.arct.cam.ac.uk/sites/www.arct.cam.ac.uk/files/p_33campbell.pdf (accessed 15 December 2024).

Cavendish E. (2024). How Does My Kitchen Sink Drain to the Sewer? Unveil the Secret! 10 Feb. https://tapron.co.uk/blogs/news/w-does-my-kitchen-sink-drain-to-the-sewer (accessed 22 March 2025).

Chaplin J. (2014). Subsidiarity and social pluralism. In: Global Perspectives on Subsidiarity. Ius Gentium: Comparative Perspectives on Law and Justice, M. Evans and A. Zimmermann (eds.), Springer, Dordrecht, Vol. **37**, pp. 72–83, https://doi.org/10.1007/978-94-017-8810-6_5; further cited by Burbridge D. (2017). p. 144

Chicago Water Tower (2014). https://en.wikipedia.org/wiki/Chicago_Water_Tower (accessed 22 March 2025).

Clayton A. (2019). San Francisco airport announces ban on sales of plastic water bottles. *Guardian*, 2 Aug 2019. https://www.theguardian.com/us-news/2019/aug/02/san-francisco-international-airport-plastic-water-bottle-ban (accessed 22 March 2025).

Clayton F. (2023). Why Namibia's 1960s sewage purifying plant is a beacon of hope for the US water crisis. *Nature Africa*, News Feature, 08 Dec 2023. https://www.nature.com/articles/d44148-023-00349-z (accessed 22 March 2025).

Corcoran E., Nellemann C., Baker E., Bos R., Osborn D. and Savelli H. (eds.), (2010). Sick Water? The central role of wastewater management in sustainable development. A Rapid Response Assessment. United Nations Environment Programme, UN-HABITAT, GRID-Arendal. https://www.grida.no/publications/218 (accessed 22 March 2025).

Council of the European Union (2024). Proposal for a Directive of the European Parliament and of the Council concerning urban wastewater treatment (recast). File: 2022/0345(COD). 1 March 2024.

Cultural heritage in action (2021). Water tower quenches cultural thirst. 23 Nov 2021. https://culturalheritageinaction.eu/water-tower-quenches-cultural-thirst/ (accessed 22 March 2025).

DANVA (Danish Water and Wastewater Association) (2022). Water in figures. 2022 Denmark. p. 64.

Digap E. (2021) Why are so many people still using plastic water bottles for travel? https://basq.livelarq.com/sustainability/why-are-so-many-people-still-using-plastic-water-bottles-for-travel/Tourismandtheplasticpollutioncrisis (accessed 15 Dec 2024).

Douglas M. C. (1966). Purity and Danger. An Analysis of the Concepts of Pollution and Taboo. Routledge, London and New York, p. 2.

Duque N., Scholten L. and Maurer M. (2024). When does infrastructure hybridisation outperform centralised infrastructure paradigms? – Exploring economic and hydraulic impacts of decentralised urban wastewater system expansion. *Water Research*, **254**, 121327, https://doi.org/10.1016/j.watres.2024.121327

Eagle Wisconsin Water Tower (2013). https://fi.wikipedia.org/wiki/Tiedosto:Eagle_Wisconsin_Water_Tower.jpg (accessed 22 March 2025).

Engelhaupt E. (2017). Huge Blobs of Fat and Trash Are Filling the World's Sewers. *National Geography*, 16 Aug 2017. https://www.nationalgeographic.com/science/article/fatbergs-fat-cities-sewers-wet-wipes-science (accessed 22 March 2025).

EPA (2023). Water Infrastructure Challenge. https://www.epa.gov/sustainable-water-infrastructure/water-infrastructure-challenge (accessed 22 March 2025).

EPA (2024). N/P Criteria Progress Map. https://www.epa.gov/nutrientpollution/state-progress-toward-adopting-numeric-nutrient-water-quality-criteria-nitrogen#tb1 (accessed 22 March 2025).

EPCA (European Chemical Agency). (n.d.). Per- and polyfluoroalkyl substances (PFAS). https://echa.europa.eu/hot-topics/perfluoroalkyl-chemicals-pfas (accessed 22 March 2025).

EurEau (2018). Time to stop the release of wet wipes into sewers. https://www.eureau. org/news/261-time-to-stop-the-release-of-wet-wipes-into-sewers (accessed 22 March 2025).

EurEau (2021). Europe's Water in Figures. An overview of the European drinking water and waste water sectors, p. 23. https://www.eureau.org/resources/publications/eureau-publications/5824-europe-s-water-in-figures-2021/file (accessed 22 March 2025).

EurEau (2025). EurEau Raises Concerns About ISO/DIS 18671 Standard. https:// www.eureau.org/news/950-eureau-raises-concerns-about-iso-dis-18671-standard (accessed 22 March 2025).

European Commission (2024). Water. https://environment.ec.europa.eu/topics/water_en (accessed 22 March 2025).

European Commission (n.d., a). Rural development. https://ec.europa.eu/regional_policy/policy/themes/rural-development_en (accessed 22 March 2025).

European Commission (n.d., b). Water Framework Directive. https://environment. ec.europa.eu/topics/water/water-framework-directive_en (accessed 22 March 2025).

Eutrophication (2024). https://en.wikipedia.org/wiki/Eutrophication (accessed 22 March 2025).

Fatberg (2024). Article. https://en.wikipedia.org/wiki/Fatberg#References (accessed 22 March 2025).

Finnish Forest Industries (2024). Forest industry production volumes since 1960, 11 Febr 2024. https://www.metsateollisuus.fi/newsroom/forest-industry-production-volumes-since-1960 (accessed 22 March 2025).

FIWA (Finnish Water Utilities Association) (2020). Taloudellisesti ja ympäristön kannalta kestävä vedenkäytön tehostaminen talousvesihuollossa Suomessa (Improving economically and environmentally sustainable water use in Finland). Monistesarja no 60, Vesilaitosyhdistys, Helsinki, Finland, p. iv.

Gleick P. (2023). The Three Ages of Water: Prehistoric Past, Imperiled Present, and a Hope for the Future. Public Affairs/Hachette, New York.

Global Handwashing Partnership (2022). 2022 Hand Hygiene Research Summary. https://globalhandwashing.org/wp-content/uploads/2023/05/HH-2022-Research-Summary_Final.pdf (accessed 22 March 2025).

Greene J. (2018). Bottled water in Mexico: The rise of a new access to water. *WIREs Water*, **5**, e1286, p. 1, https://doi.org/10.1002/wat2.1286

Grewar T. (2019). South Africa's options for mine impacted water re-use: a review. *The Journal of the South African Institute of Mining and Metallurgy*, **119**, 321–331.

Grigg N. S. (2024). 6 April 2024, Personal communication, Colorado State University.

Grigg N. S., Rogers P. D. and Edmiston S. (2011). Retrospective analysis of performance of dual distribution systems. Web Report #4333, Water Research Foundation, Denver, Colorado, USA. https://www.waterrf.org/system/files/resource/2019-07/INFR1SG09b-4333_0.pdf (accessed 22 March 2025).

Haapalainen P. (1996). (Original in Finnish) Subsidiariteettiperiaatteen merkitys suomalaisen yhteiskunnan kehittämisessä. The significance of the subsidiarity principle for the development of the Finnish society. MSc thesis, Department of Regional Studies, University of Tampere, p. 9. https://trepo.tuni.fi/handle/10024/84304 (accessed 22 March 2025).

Haarhoff J. and Van der Merwe B. (1996). Twenty-five years of wastewater reclamation in Windhoek, Namibia. *Water Science and Technology*, **33**(10–11), 25–35, https://doi.org/10.2166/wst.1996.0658

Heikkila T. (2004). Institutional boundaries and common-pool resource management: a comparative analysis of water management programs in California. *Journal of Policy Analysis and Management*, **23**(1), 97–117, https://doi.org/10.1002/pam.10181

Heino O. and Takala A. (2013). Halpaa eli hyvää – minkälaisia merkityksiä vesihuoltoala rakentaa itsestään (cheap and good – what kind of significations does water services sector build of itself?). *Kunnallistieteellinen Aikakauskirja*, **41**(3), 226–245.

History Timelines. A History Timeline About Hamburg. https://historytimelines.co/timeline/hamburg (accessed 22 March 2025).

Hooghe L. and Marks G. (2002). Types of Multi-Level Governance. *Les Cahiers européens de Sciences Po*, n° 03. https://www.sciencespo.fr/centre-etudes-europeennes/sites/sciencespo.fr.centre-etudes-europeennes/files/n3_2002_final.pdf (accessed 22 March 2025).

Hörkkö E. (2023). 23 Oct 2023, Personal communication, Finnish Water Utilities Association.

Hubbart S. and Ross L. (2024). The Increasing Demand and Decreasing Supply of Water. National Environmental Education Foundation. https://www.neefusa.org/story/water/increasing-demand-and-decreasing-supply-water (accessed 22 March 2025).

Hukka J. J. and Katko T. S. (2015a). Appropriate pricing policy needed worldwide for improving water services infrastructure. *Journal AWWA*, **107**(1), E37–E46, https://doi.org/10.5942/jawwa.2015.107.0007

Hukka J. J. and Katko T. S. (2015b). Resilient asset management and governance for deteriorating water services infrastructure. 8th nordic conference on construction economics and organization. *Procedia Economics and Finance*, **21**, 112 119. http://www.sciencedirect.com/science/journal/22125671/21 (accessed 22 March 2025), https://doi.org/10.1016/S2212-5671(15)00157-4

Ilemobade A. A., Adewumi J. R. and van Zyl J. E. (2009). Assessment of the Feasibility of Using a Dual Water Reticulation System in South Africa. Report to the Water Research Commission. https://www.wrc.org.za/wp-content/uploads/mdocs/1701-1-091.pdf (accessed 22 March 2025).

Immerzeel W. W., Lutz A. F., Andrade M., Bah A., Biemans H., Bolch T., Hyde S., Brumby S., Davies B. J., Elmore A. C., Emmer A., Feng M., Fernández A., Haritashya U., Kargel J. S., Koppes M., Kraaijenbrink P. D. A., Kulkarni A. V., Mayewski P. A., Nepal S., Pacheco P., Painter T. H., Pellicciotti F., Rajaram H., Rupper S., Sinisalo A., Shrestha A. B., Viviroli D., Wada Y., Xiao C., Yao T. and Baillie J. E. M. (2020). Importance and vulnerability of the world's water towers. *Nature*, **577**, 364–369, https://doi.org/10.1038/s41586-019-1822-y

Insider Monkey (2023). 5 Largest Biodiesel Producers in the World, 8 Sept 2023. https://www.insidermonkey.com/blog/5-largest-biodiesel-producers-in-the-world-1191630/5/ (accessed 22 March 2025).

IWA (2023). 1st IWA Non-Sewered Sanitation Conference. https://nssconference.org/ (accessed 22 March 2025).

IWA Source (2016). Corroded pipes, eroded trust. Debate, May 2016, 69–73.

Jacobs S. and Ladegaard P. (2010). Regulatory Governance in Developing Countries. Better Regulation for Growth: Governance Frameworks and Tools for Effective Regulatory Reform. ICAS (Investment Climate Advisory Services, World Bank Group), p. 7. http://regulatoryreform.com/wp-content/uploads/2014/11/Regulatory-Governance-Jacobs-Ladegaard-2010.pdf (accessed 22 March 2025).

Joksimovic D., Khan A. and Orr B. (2020). Inappropriate disposal of 'flushable' consumer products – reasons for concern. *Water Science & Technology*, **81**(1), 102–108, https://doi.org/10.2166/wst.2020.087

Jones E. R., van Vliet M. T. H., Qadir M. and Bierkens M. F. P. (2021). Country-level and gridded estimates of wastewater production, collection, treatment and reuse. *Earth System Science Data*, **13**, 237–254, p. 237, https://doi.org/10.5194/essd-13-237-2021

Juuti P. and Katko T. (1998). Ernomane vesitehdas – Tampereen vesilaitos 1835–1998 (Marvellous water factory – Tampere Water Works 1835–1998. Tampere City Water

Works, Tampere, Finland, p. 137. (In Finnish, Summary in English and Swedish). https://trepo.tuni.fi/handle/10024/66324 (accessed 22 March 2025).

Juuti P. and Katko T. (eds.), (2005). Water, Time and European cities. History matters for the futures. Printed in EU. http://www.watertime.net/ (accessed 22 March 2025).

Juuti P. and Wallenius K. (2005). Kaivot ja käymälät. Brief History of Wells and Toilets. KehräMedia Inc., Kangasala. (In Finnish and English). https://trepo.tuni.fi/handle/10024/65357 (accessed 22 March 2025).

Juuti P., Katko T., Persson K. and Rajala R. (2009). Shared history of water supply and sanitation in Finland and Sweden, 1860–2000. *Vatten*, **65**(3), 165–175.

Kadri S. U. T., Tavanappanavar A. N., Babu R. N., Bilal M., Singh B., Gupta S. K., Bharagava R. N., Govarthanan M., Savanur M. A. and Mulla S. I. (2021). Overview of waste stabilization ponds in developing countries. In: Cost-efficient Wastewater Treatment Technologies, The Handbook of Environmental Chemistry, M. Nasr and A. M. Negm (eds.), Springer, Cham, Germany, Vol. **117**, pp. 153–175, https://doi.org/10.1007/698_2021_790

Kallio J. (2020). Jätevesineuvonta haja-asutusalueilla 2011–2019 – Loppuraportti (Abstract: On-site wastewater treatment guidance 2011–2019, p. 5. Suomen ympäristökeskuksen raportteja; 48/2020, Helsinki, Finland. https://vesi.fi/aineistopankki/jatevesineuvonta-haja-asutusalueilla-2011-2019-loppuraportti/ (accessed 22 March 2025).

Kallioinen S. (2000). Kaksivesijärjestelmä vesihuollon osana (Dual system in water services). Lisensiaatintyö, Rakennustekniikan osasto, Tampereen teknillinen korkeakoulu, Finland.

Kaloinen J. (2015). 30 March 2015, Personal communication.

Katko T. (unpublished 2013). NSS mission polytechnic of Namibia project: UWAS matkakertomus 8–21 Sept 2013.

Katko T. S. (2016). Finnish Water Services – Experiences in Global Perspective. Finnish Water Utilities Association, Helsinki, Finland. Co-published E-book, IWA Publishing, London, UK, 2017. https://iwaponline.com/ebooks/book/328/Finnish-Water-Services-Experiences-in-Global?redirectedFrom=PDF (accessed 22 March 2025).

Katko T. S. and Hukka J. J. (2015). Social and economic importance of water services in the built environment: need for more structured thinking. 8th nordic conference on construction economics and organization. Elsevier. *Procedia Economics and Finance*, **21**, 217–223. p. 221, https://doi.org/10.1016/S2212-5671(15)00170-7

Katko T. and Juuti S. (2007). Watering the city of Tampere from the mid-1800s to the 21st century. Tampere Water and International Water History Association, p. 28. https://tampub.uta.fi/handle/10024/65709 (accessed 15 December 2024).

Kemira (2020). Highlights from an international consumer survey on water: consumers' views on the value of water, their own water use, and concerns and responsibilities related to water supply. https://www.kemira.com/app/uploads/2020/10/Kemira_water_datasummary_US_FINAL-5f9bf7b272098.pdf (accessed 22 March 2025).

Knuuttila S. (2024). 23 Sept 2024, Personal communication, Helsinki, Finland.

Köppä T. (2022). Concern for community: contributions of the cooperatives and fraternité to sustainable development. In: Perspectives on Cooperative Law, W. Tadjudje and I. Douvitsa (eds.), Springer Nature, Singapore, pp. 211–222.

Korhonen A., Kuusela M., Liski-Markkanen S. and Marjomaa T. (2020) Kestävä veden käyttö – vedenkäyttöselvitys (Sustainable water use – a survey on water use). Julkaisu 453., Työtehoseura, Helsinki, Finland.

Kurki V. (2010) Unpublished survey. 8 May 2010, n = 38.

Kuulas A., Renko T. and Kuivamäki R. (2020). Investment needs in the water service sector in Finland by 2040. Publication series no. 63, Finnish Water Utilities

Association, Helsinki. https://www.vvy.fi/site/assets/files/5239/vesihuollon_investointitarpeet_vvy_10092020_final.pdf (accessed 15 December 2024).

Kyiv Water Museum (n.d.). https://en.wikipedia.org/wiki/Kyiv_Water_Museum (accessed 22 March 2025).

Kytö J. (1974). Teollisuuden vedenhankinta ja veden käyttö (Industrial water abstraction and use), Tiedotus 71, Vesihallitus, Helsinki, p. 144.

Lapinlampi T. (2021). Vesihuoltolaitokset 1970–2014 (Water utilities 1970–2014). Suomen ympäristökeskuksen raportteja 21/2021. https://vesi.fi/aineistopankki/wp-content/uploads/2022/11/SYKEra_21_2021_Vesihuoltolaitokset-1970-2014.pdf (accessed 22 March 2025).

Lappalainen K. M. (2018). A renewed diagnosis and paradigm for eutrophication of the Baltic Sea. Acta Univ. Oul. C 663. Doctoral dissertation, Oulu university (Abstract in English) https://oulurepo.oulu.fi/bitstream/handle/10024/36270/isbn978-952-62-1941-7.pdf?sequence=1&isAllowed=y (accessed 22 March 2025).

Laugeri L. (1987). Water for all – who pays? *World Health Forum*, **8**, 453–460.

Laukka V., Kallio J., Herrmann I., Malila R., Nilivaara R. and Heiderscheidt E. (2022). Governance of on-site sanitation in Finland, Sweden and Norway. Reports of the Finnish Environment Institute 8, 2022, p. 3. https://helda.helsinki.fi/items/b1c850e3-49f7-47b1-8d79-e5f40e8d767d (accessed 22 March 2025).

Lloyd R. (2022). Replacing Lead Water Pipes with Plastic Could Raise New Safety Issues. *Scientific American*, 16 Aug 2022. https://www.scientificamerican.com/article/replacing-lead-water-pipes-with-plastic-could-raise-new-safety-issues/ (accessed 22 March 2025).

Louisville Water Tower (2024). https://en.wikipedia.org/wiki/Louisville_Water_Tower (accessed 22 March 2025).

Low B., Ostrom E., Simon C. and Wilson J. (2003). Redundancy and diversity: do they influence optimal management? In: Navigating Social-Ecological Systems: Building Resilience for Complexity and Change, F. Berkes, J. Colding and C. Folke (eds.), Cambridge University Press, p. 108.

Makkonen A. (2020). Pakatun veden tuotanto ja haasteet – keskeiset toimijat ja viennin tilanne (Production and Challenges of Packaged Water – major Players and State of Exporting). Ab Picus Advisors Oy, Helsinki, p. 3.

Marine Finland. Internal loading regulates marine eutrophication. https://www.marinefinland.fi/enUS/Nature_and_how_it_changes/State_of_the_Baltic_Sea/Eutrophication/Internal_loading (accessed 15 December 2024).

Mason S. A., Welch V. G. and Neratko J. (2018). Synthetic polymer contamination in bottled water. *Frontiers in Chemistry*, **6**, 407, https://doi.org/10.3389/fchem.2018.00407

Mattila H. (2005). Appropriate management of on-site sanitation. Doctoral dissertation, Tampere University of Technogy, Publications, Vol. **537**, p. 94. https://trepo.tuni.fi//handle/10024/114828 (accessed 22 March 2025).

Mattila H. (2024). 13 Nov 2024, Personal communication, Finland.

Maxcy-Brown J., Elliott M. A. and Bearden B. (2023). Household level wastewater man and disposal data USA. *Water Policy*, **25**(9), 927–947, https://doi.org/10.2166/wp.2023.147

McDonald H. (2014). Introduction of water charges in Ireland cause widespread street protests. *Guardian*, 31 Oct 2014. https://www.theguardian.com/world/2014/oct/31/water-charges-ireland-cause-protests (accessed 22 March 2025).

Merton (n.d.). Disposal of fats and oils. https://www.merton.gov.uk/business-and-consumers/food-safety/disposal-of-fats-and-oils (accessed 22 March 2025)

Michelson J. (2023). 2023 -The Year of Infrastructure Reinvention. *Forbes*, 29 Dec 2023. https://www.forbes.com/sites/joanmichelson2/2023/12/26/2023the-year-of-infrastructure-wins/ (accessed 22 March 2025).

Miettinen I. T. and Pursiainen A. (2009). Pakattujen vesien laadun tarkastelua (assessing the quality of *packaged* waters). *Ympäristö ja Terveys-lehti*, **40**(3), 34–36.

Murdoch J. H. (1956). 75 Years of too cheap water. *Journal AWWA*, **48**(8), 925–930, https://doi.org/10.1002/j.1551-8833.1956.tb20304.x

NACWA (National Association of Clean Water Agencies). Toilets Are Not Trashcans. https://www.nacwa.org/advocacy-analysis/campaigns/toilets-are-not-trashcans (accessed 22 March 2025).

Nagler B. E. (1966). Elevated concrete reservoirs in Finland. *Journal AWWA*, **58**(11), 1429–1445, https://doi.org/10.1002/j.1551-8833.1966.tb01712.x

Najar N. and Persson K. M. (2023). Status improvement in water and wastewater fixed facilities: success and challenges of 11 Swedish water utilities as case studies. *Water Policy*, **25**(7), 656–679, https://doi.org/10.2166/wp.2023.263

Neste (2023). Circular-economy, 12 Dec 2023. https://www.neste.com/fi-fi/news/neste-on-mukana-valtakunnallisessa-kinkkutemppu-kiertotaloustempauksessa-paistinrasvan-keraeys-kaeynnistyy-18-12-2023 (accessed 22 March 2025).

Oakerson R. J. (1999). Governing Local Public Economies: Creating the Civic Metropolis. Oakland, CA: ICS Press.

OECD (Organisation for Economic Co-Operation and Development (2011). Benefits of Investing in Water and Sanitation: An OECD Perspective. OECD Publishing, Paris, France, p. 16, https://doi.org/10.1787/22245081

Oras (n.d.). https://www.oras.com/en/about-oras (accessed 22 March 2025).

O'Riordan T. (2024). Review of the three ages of water: prehistoric past, imperiled present, and a hope for the future. *Environment: Science and Policy for Sustainable Development*, **66**(1), 46–47, https://doi.org/10.1080/00139157.2023.2269048

Ostrom E. (2010). Beyond markets and states: polycentric governance of Complex economic systems. *American Economic Review*, **100**, 641–672, p. 653, https://doi.org/10.1257/aer.100.3.641

Ostrom V. and Ostrom E. (1991). Public goods and public choices: the emergence of public economies and industry structures. In: The Meaning of American Federalism, V. Ostrom (ed.), Institute for Contemporary Studies Press, San Francisco, pp. 163–197.

Ostrom E., Schroeder L. and Wynne S. (1993). Institutional Incentives and Sustainable Development: Infrastructure Policies in Perspective. Westview Press Inc., Boulder, Colorado.

Papunen P. (1986). Monitasosuunnittelu alueellisessa kehittämisessä – alhaalta ylöspäin suuntautuvan kehittämisen mahdollisuuksista aluepoliittisen suunnittelujärjestelmän toiminnassa. (Multi-level planning in regional development – on the possibilities of the bottom-up approach development in the regional planning system). Research Reports B 38/1986. Department of Regional Studies, University of Tampere, Finland, p. 52.

Parag Y. and Roberts J. T. (2009). A battle against the bottles: building, claiming, and regaining tap water trustworthiness. *Society & Natural Resources*, **22**(7), 625–636, p. 625, https://doi.org/10.1080/08941920802017248

Parks R. B. and Oakerson R. J. (2000). Regionalism, localism, and metropolitan governance: suggestions from the research program on local public economies. *State and Local Government Review*, **32**(3), 169–179, https://doi.org/10.1177/0160323X0003200302

Persson K. M. (2024). 4 Nov 2024, Personal communication, Lund University, Sweden.

Pinter G., Travnicek L. and Arbeiter F. (2024). 100 years lifetime of plastic pipes. Commissioned by the European Plastic Pipes and Fittings Association. https://www.teppfa.eu/wp-content/uploads/2024-04-25-Meta-study-100-years-of-lifetime-of-plastic-pipes.pdf (accessed 22 March 2025).

Prasetiawan T., Nastiti A. and Muntalif B. S. (2017). 'Bad' piped water and other perceptual drivers of bottled water consumption in Indonesia. *WIREs Water*, **4**, e1219, p. 1, https://doi.org/10.1002/wat2.1219

ProCon.org (2021). Pro and Con: Bottled Water Ban. https://www.britannica.com/story/pro-and-con-bottled-water-ban (accessed 22 March 2025).

Rajala R. P. and Katko T. S. (2004). Household water consumption and demand management in Finland. *Urban Water Journal*, **1**(1), 17–26, https://doi.org/10.1080/15730620410001732080

Ridder M. (2023). Bottled water market worldwide – statistics & facts. 8 Aug 2023. https://www.statista.com/topics/9561/bottled-water-market-worldwide/#topicOverview (accessed 22 March 2025).

Rodwan J. G.Jr. (2019). Significant but slower growth for bottled water in 2018. *BWR* June/July 2019. p. 17. https://bottledwater.org/wp-content/uploads/2020/03/2018 BottledWaterStats_pub2019.pdf (accessed 22 March 2025).

Rosinger A. Y. (2022). Using water intake dietary recall data to provide a window into US water insecurity. *The Journal of Nutrition*, **152**, 1263–1273, https://doi.org/10.1093/jn/nxac017

Rosinski A. (2014). How Much Grease Fast Food Places Put Out. *Greener ideal*, 27 Sept 2014. https://greenerideal.com/news/business/0624-restaurant-grease-being-stolen/ (accessed 22 March 2025).

RWSN Executive Steering Committee (2010). Myths of the Rural Water Supply Sector. Rural Water Supply Network Perspectives No 4. https://www.ircwash.org/sites/default/files/RWSN-2010-Myths.pdf (accessed 22 March 2025).

Salmi H. (2017). Catastrophe, emotions and guilt – The great fire of Turku 1827. In: Catastrophe, Gender and Urban Experience, 1648–1920, D. Simonton and H. Salmi (eds.), Routledge, pp. 121–138.

Schaverien A. (2019). Scientists Solve a Puzzle: What's Really in a Fatberg. *New York Times*, 4 Oct 2019. https://www.nytimes.com/2019/10/04/world/europe/sidmouth-fatberg.html (accessed 22 March 2025).

Schindler D. W. (2012). The dilemma of controlling cultural eutrophication of lakes. *Proceedings of the Royal Society B: Biological Sciences*, **279**, 4322–4333, https://doi.org/10.1098/rspb.2012.1032

Schindler D. W., Hecky R. E., Findlay D. L., Stainton M. P., Parker B. R., Paterson M. J., Beaty K. B., Lyng M. and Kasian S. E. M. (2008). Eutrophication of lakes cannot be controlled by reducing nitrogen input: results of a 37-year whole-ecosystem experiment. *Proceedings of the National Academy of Sciences*, **105**(32), 11254–11258, https://doi.org/10.1073/pnas.0805108105

Sedlak D. (2014). Water 4.0. The Past, Present and Future of the World´S Vital Resource. Yale University Press, New Haven, Connecticut, p. 87.

Stadtfeld O. R. and Schlaweck K. L. (1988). International comparison of water prices. *Aqua*, **37**(4), 173–177.

Stenstedt L. (2019). The Asbestos Beneath Our Streets. *Water World*, 1 Feb 2019. https://www.waterworld.com/drinking-water-treatment/distribution/article/16190948/the-asbestos-beneath-our-streets (accessed 22 March 2025).

Stoa R. (2014). Subsidiarity in principle: decentralization of water resources management. *Utrecht Law Review*, **10**, 31–, https://doi.org/10.18352/ulr.267

St. Petersburg, Museum of Water (n.d.). http://www.saint-petersburg.com/museums/museum-water/ (accessed 22 March 2025).

Sunela M. (2024). 2 March 2024, Personal communication, Fluidit. https://fluidit.com/

Sunela M. I. and Puust R. (2017). Real-time whole-cost optimization of water production and distribution. CCWI (Control and Computing for the Water Industry)

–Conference, 5–7 Sept 2017, University of Sheffield, England. https://trepo.tuni.fi/handle/10024/129384 (accessed 15 December 2024).

Suominen A. (2024). 5 Nov 2024, Personal communication, Finland.

Sustainable sanitation alliance. https://www.susana.org/en/knowledge-hub/trainings-conference-and-events-materials/conferences/2018/733-dry-toilet-conference-2018# (accessed 15 December 2024).

Sutton S. and Butterworth J. (2021). Self-Supply: Filling the Gaps in Public Water Supply Provision. Practical Action Publishing, Rugby, UK, https://doi.org/10.3362/9781780448190

SYKE (Finnish Environment Institute) (2021). Vesistöjen kuormitus ja luonnon huuhtouma (Loadings of water bodies and natural discharge). 27 Oct 2021. https://www.ymparisto.fi/en/state-environment/water/nutrient-load (accessed 22 March 2025).

Sänger N., Heinzel C. and Sandholz S. (2021). Advancing resilience of critical health infrastructures to cascading impacts of water supply outages – insights from a systematic literature review. *Infrastructures* **6**, 177, p. 1, https://doi.org/10.3390/infrastructures6120177

Takala A. (2017). Understanding sustainable development in Finnish water supply and sanitation services. *International Journal of Sustainable Built Environment*, **6**, 501–512, https://doi.org/10.1016/j.ijsbe.2017.10.002

Tamminen T. and Andersen T. (2007). Seasonal phytoplankton nutrient limitation patterns as revealed by bioassays over Baltic Sea gradients of salinity and eutrophication. *Marine Ecology Progress Series*, **340**, 121–138, https://doi.org/10.3354/meps340121

Taylor M. (2017). Total monster': fatberg blocks London sewage system. *Guardian*, 12 Sept 2017. https://www.theguardian.com/environment/2017/sep/12/total-monster-concrete-fatberg-blocks-london-sewage-system (accessed 22 March 2025).

Tebbutt G. (2016). A European Traveller's Guide To Drinking Water. Citybase Apartments, 10 June 2016 [Infographic] https://www.citybaseapartments.com/blog/a-european-travellers-guide-to-drinking-water-infographic/ (accessed 22 March 2025).

The Association of Consulting Engineering Companies Canada, the Canadian Construction Association, the Canadian Parks and Recreation Association, the Canadian Public Works Association, the Canadian Society for Civil Engineering, the Canadian Urban Transit Association, the Canadian Network of Asset Managers, and the Federation of Canadian Municipalities (2019). Canadian infrastructure report card. http://canadianinfrastructure.ca/downloads/canadian-infrastructure-report-card-2019.pdf (accessed 22 March 2025).

The Helsinki Term Bank for the Arts and Science (2021). Virtue ethics. (Original in Finnish) https://tieteentermipankki.fi/wiki/Filosofia:hyve-etiikka (accessed 22 March 2025).

The Mercury News (2007). CA builders OK to use plastic water pipe. 30 Jan 2007.

UN (n.d.). Sustainable Development Goals. Communications materials. https://www.un.org/sustainabledevelopment/water-and-sanitation/ (accessed 22 March 2025).

UN Habitat and WHO (2021). Progress on wastewater treatment – Global status and acceleration needs for SDG indicator 6.3.1. Geneva, Switzerland. https://unhabitat.org/progress-on-wastewater-treatment-%E2%80%93-2021-update (accessed 22 March 2025).

UNU (United Nations University) (2023). Bottled Water Masks World's Failure to Supply Safe Water for All. 16 Mar 2023. Press Release. https://unu.edu/press-release/bottled-water-masks-worlds-failure-supply-safe-water-all (accessed 22 March 2025).

Uusi Suomi (1963). Veden käytön jatkuvalla kasvulla ei ole ylärajaa (The continuous growth of water use will have no limit), 11 Nov 1963.

Wainaina G. K. and Barbosa H. (2024). Revisiting the self-supply model: A foundational, not a complimentary model water supply in rural households. *World Water Policy*, **10**(4), 1162–1169, p. 1, https://doi.org/10.1002/wwp2.12222

Walker T. (2018). Microplastics in bottled water. https://www.dw.com/en/bottled-water-not-safe-from-microplastic-contamination/a-42936246 (accessed 22 March 2025).

Wallace T., Gibbons D., O'Dwyer M. and Curran T. P. (2017). International evolution of fat, oil and grease (FOG) waste management–A review. *Journal of Environmental Management*, **187**, 424–435, p. 424, https://doi.org/10.1016/j.jenvman.2016.11.003

Water Museums Global Network (n.d.). https://www.watermuseums.net/ (accessed 22 March 2025).

Water tower (n.d.). https://en.wikipedia.org/wiki/Water_tower (accessed 22 March 2025).

Wavin (2024). How to comply with new EU law on lead in drinking water. 18 June 2024. https://blog.wavin.com/en-ie/how-to-comply-with-new-eu-law-on-lead-in-drinking-water (accessed 22 March 2025).

WHO, UNICEF and World Bank (2022). State of the World's Drinking Water: an Urgent Call to Action to Accelerate Progress on Ensuring Safe Drinking Water for All. World Health Organization, Geneva, Switzerland. https://www.who.int/publications/i/item/9789240060807 (accessed 22 March 2025).

Wolf J. R. (2020), Water Towers: Iconic Infrastructure, Underutilized Opportunity. *Common Edge*, 23 Nov 2020. https://commonedge.org/water-towers-iconic-infrastructure-underutilized-opportunity/ (accessed 22 March 2025).

Wolf J., Hubbard S., Wolf J., Hubbard S., Braue r M., Ambelu A., Arnold B. F., Bain R., Bauza V., Brown J., Caruso B. A., Clasen T., Colford JrJ. M., Freeman M. C., Gordon B., Johnston R. B., Mertens A., Prüss-Ustün A., Ross I., Stanaway J., Zhao J. T., Cumming O. and Boisson S. (2022). Effectiveness of interventions to improve drinking water, sanitation, and handwashing with soap on risk of diarrhoeal disease in children in low-income and middle-income settings: a systematic review and meta-analysis. *The Lancet*, **400**, 48–59, https://doi.org/10.1016/S0140-6736(22)00937-0

WWAP (United Nations World Water Assessment Programme) (2017). The United Nations World Water Development Report 2017. Wastewater: The Untapped Resource. UNESCO, Paris, France, p. 2. https://www.unep.org/resources/publication/2017-un-world-water-development-report-wastewater-untapped-resource (accessed 22 March 2025).

Yakel D. (2021). SFO Expands Plastic-Free Policy to Prohibit Sale of All Plastic Bottles. 5 April 2021. https://www.flysfo.com/media/press-releases/sfo-expands-plastic-free-policy-prohibit-sale-all-plastic-bottles (accessed 22 March 2025).

YLE (2009). Finland Beats EU Commission in Waste Water Dispute, 6 Oct 2009. https://yle.fi/a/3-5894023 (accessed 22 March 2025).

Yle News (2018). Finland pushes tap water at World Junior Athletics Championships, 13 July 2018. https://yle.fi/a/3-10303369 (accessed 22 March 2025).

Zimmermann A. (2022). Brussels holds its nose on sewage water pollution problem. *Politico*, 24 Nov 2022. https://www.politico.eu/article/brussels-holds-its-nose-on-sewage-water-pollution-problem/ (accessed 22 March 2025).

doi: 10.2166/9781789064162_0087

Chapter 4

Rules of water services: how to promote teamwork?

'We never know the worth of water until the well is dry.' (English Proverb)

'Plans fail for lack of counsel, but with many advisers they succeed.' (Proverbs 15:22)

4.1 INTRODUCTION

Numerous studies have addressed the relevance of water utility performance globally. These studies point out *the importance of the institutional factors* affecting those managing water utilities and those providing regulatory oversight: social structures (the political and cultural context), formal organisations (government ministries and regulatory authorities), and support systems (including political patronage and civil service). These external factors affect how conflicts are resolved regarding resource allocation, pricing and access to water services. Obviously, these issues influence the internal governance of water utilities (Berg, 2013).

The most popular definitions of an institution have at their core social factors that influence, to some extent, human behaviour (Davis, 2009). According to North (1993, pp. 5–7), institutions are the 'rules of the game of a society, or more formally, are the humanly devised constraints that shape human interaction'. They are composed of formal rules (statute law, common law, regulations), informal constraints (conventions, norms of behaviour and self-imposed codes of conduct), and the enforcement characteristics of both. Together with the standard constraints of economics, institutions define the choice set and, therefore, determine transaction and production costs. They evolve incrementally, connecting the past with the present and the future. The admixture of rules, norms and enforcement characteristics determines economic performance. Hendricks (2010) defined institutions as society's means to govern itself.

North (1993, p. 2) explained that institutions are formed to reduce uncertainty in human exchange. According to North (1993, p. 5) it shall also be of the utmost importance to distinguish clearly institutions from organisations. The government authorities, provincial administration, water utilities, enterprises and associations, and in some instances the individuals, can be defined as organisations. These definitions support five propositions that define the essential characteristics of institutional change (North 1993, p. 6):

(1) The continuous interaction of institutions and organisations in the economic setting of scarcity so that competition is the key to institutional change.
(2) Competition forces organisations to continually invest in skills and knowledge to survive. The kinds of skills and knowledge acquired by individuals and their organisations will shape evolving perceptions of opportunities and hence choices that will incrementally alter institutions.
(3) The institutional framework dictates the kinds of skills and knowledge perceived to have the maximum pay-off.
(4) Perceptions are derived from the mental constructs of the players.
(5) The economies of scope, complementarities, and network externalities of an institutional matrix make institutional change overwhelmingly incremental and path-dependent.

The water services playground

North (1994) explained that organisations are functioning in such a way that they try to utilise available opportunities within the boundaries of the institutional framework. North (1990) compared the institutions with the rules of competitive team sports, which consist of formal and written rules and underlying and supplementary unwritten codes of conduct. If the rules or informal codes are violated, some sort of punishment or penalty will often be enacted. An essential part of the functioning of institutions is the high cost of ascertaining violations and severity of punishment.

Using the 'soccer analogy' of North (2003), Figure 4.1 shows an example on how utility-based water services are played in Finland where water services are provided by the local governments (municipalities) and produced by municipal utilities or water cooperatives. As in good soccer, we need skillful players (competent staff), good team play (good governance), proper training and coaching (education, capacity building), active supporters (fans, citizens, media), managers (policymakers), referees (permit authorities), rules and sidelines on the field (legislation), and a playground (infrastructure) (Inha *et al.*, 2019a, p. 1).

In addition, we need supporting services such as physiotherapy for recovery (operation and maintenance). Water sector professionals as well as being very good players have the ability to alter their roles, making career changes and widening their views and competence. Besides, they have a good eye for the

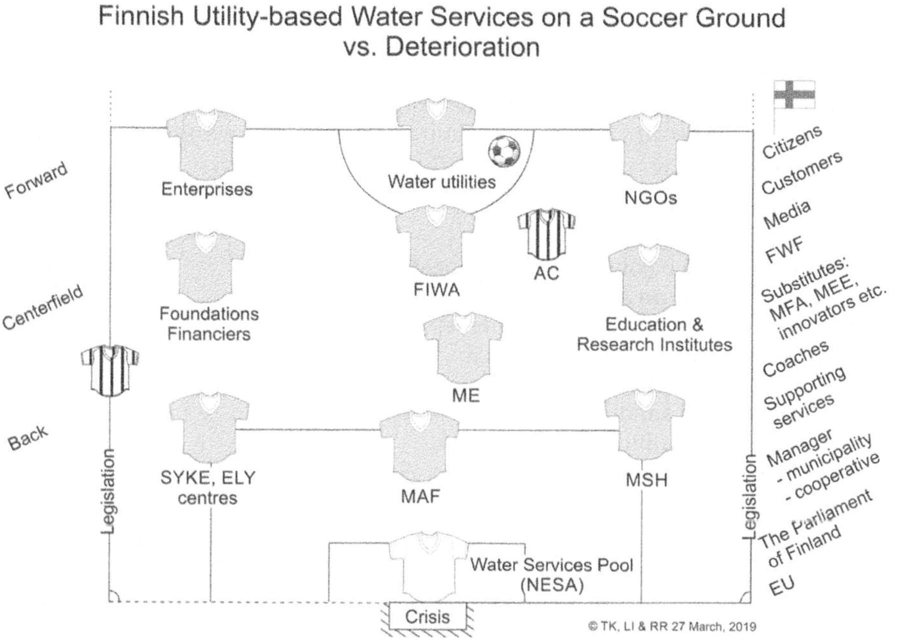

Figure 4.1 Finnish soccer ground for utility-based water services.
Source: Inha *et al.* (2019a, p. 4).

game and capacity to react accordingly. In other words, they are agile and resilient. Even a world-class team should not rely too much on one player, but on smooth collaboration and team play (Inha *et al.*, 2019a, p. 3).

According to other Finnish institutional arrangements of on-site systems, cooperatives and various forms of supramunicipal systems, the players and their roles change accordingly (Inha *et al.*, 2019a, pp. 3, 7). The smallest water producers are the on-site self-supply systems, typically serving one or a few households where the real estate owner is in charge. Small rural systems serving a village or slightly larger area, are commonly managed by Water User Associations (cooperatives), which have recently become increasingly involved also in wastewater management. In these cooperatives, a 'champion' (Katko, 1994) plays in the midfield with the municipality acting as the goalkeeper. In smaller systems the role of citizens is remarkably important. In the case of supramunicipal systems, municipal utilities play in the frontline while the manager and the board stay more in the midfield. Often, the Supreme Administrative Court acts as the goalkeeper, at least in contentious cases.

While the principles of good teamwork in soccer can be useful for describing the nature of water services, they do not apply universally. The business principles commonly found in high-level international soccer do not align with water services, which are designed to provide essential services to all citizens (Myth 14). At the same time, it is important to incorporate sound economic thinking and ensure the safeguarding of adequate investments for future infrastructure replacement–an increasingly challenging task worldwide (see introduction to chapter four).

Ostrom (2010, p. 653) listed eight design principles or best practices (as developed by Cox *et al.*, 2009) which characterise the following long-sustained institutional regimes:

(1A) User boundaries.
(1B) Resource boundaries.
(2A) Congruence (similarity) with local conditions.
(2B) Appropriation and provision.
(3)   Collective choice arrangements.
(4A) Monitoring users: individuals.
(4B) Monitoring the resource.
(5)   Graduated sanctions.
(6)   Conflict resolution mechanisms.
(7)   Minimal recognition of rights.
(8)   Nested enterprises.

These principles emphasise strongly the importance of *locality, collaboration and local democracy*. Something that works in one place and in one community does not fit necessarily into another place or community. The use of community-governed resource can have different kinds of arrangements depending on the culture and nature of the local community.

According to Scott (2008, p. 40; cited by Hatakka, 2016, p. 15), institutions can be defined as *regulative, normative and cultural-cognitive* elements that, together with the associated activities and resources, provide stability and meaning in societal life. Scott (2008, pp. 50–70, cited by Hatakka, 2016, p. 16) gives three pillars of institutions in Table 4.1.

Table 4.1 Three pillars of institutions by Scott (2008, p. 51, cited by Hatakka, 2016, p. 16).

	Regulative	**Normative**	**Cultural-Cognitive**
Basis of compliance	Expedience	Social obligation	Taken-for-grantedness Shared understanding
Basis of order	Regulative rules	Binding expectations	Constitutive schema
Mechanism	Coercive	Normative	Mimetic
Logic	Instrumentality	Appropriateness	Orthodoxy
Indicators	Rules Laws Sanctions	Certification Accreditation	Common beliefs Shared logics of action Isomorphism
Affect	Fear/Guilt/ Innocence	Shame/Honour	Certainty/Confusion
Basis of legitimacy	Legally sanctioned	Morally governed	Comprehensible Recognisable Culturally supported

A regulative approach is probably the most common and easiest way to perceive institutions. Its objective as a guiding process is to shape people's behaviour by setting constraints, monitoring and controlling compliance, and as necessary, influencing future behaviour by using coercion or threat of punishment exercised by a higher authority, or by giving incentives (Scott, 2008, pp. 50–70, cited by Hatakka, 2016, p. 16).

The normative pillar in institutional theory is built upon the interpretations of scholars and theorists who emphasise values (what is preferred or desired) and standards (how something must be done). Normative systems provide the moral and practical relevance by defining goals and objectives but also by designating appropriate ways to strive for them (Scott, 2008, p. 54, cited by Hatakka, 2016, p. 18).

The cultural-cognitive approach stresses the centrality of cultural-cognitive elements of institutions, for example the shared conceptions that constitute the nature of social reality and create the frames through which the meaning is made (Scott, 2008, p. 54, cited by Hatakka, 2016, p. 18).

However, Hatakka (2016, p. 16) understood that Scott's aim is that this classification should not be assessed as three mutually exclusive categories but rather as three different kinds of approach, which have their own distinctive characteristics. When the subject of study is examined from each of these approaches in turn, a more comprehensive and holistic understanding may be obtained than by focusing only on one approach or by limiting the focus of some approaches.

When future developments are being envisaged and outlined, various changes can be divided into the following three categories: structural changes, procedural changes and changes in attitudes. In addition, these changes must be systemically preferred and culturally accepted (Mannermaa, 1993). While the formal rules can be changed overnight, the informal norms, according to North (1993, p. 7), change only gradually.

Parviainen (2019) underlined that in political decision making it is of the utmost importance that the informal institutions can be identified and are recognised, and the motives behind them and their possible consequences are understood thoroughly.

The iceberg metaphor is used by Andrews (2013) to represent that 'a large part of institutional logic is unseen or below the water line because it is informal'. He further pointed out that institutional reforms can only work if they are tailored to the local context and, therefore, the so-called best practice reforms tend to fail. Hence, according to Andrews (2013), new institutions can be created that resemble iceberg tips with no foundation (Figure 4.2).

The tip seems to be intact above the waterline, but due to the lack of norms and cultural-cognitive fundamentals, it is only a matter of time until it sinks (Andrews, 2013). We should also recognise that informal institutions are produced internally, that is they are endogenous to a community (Lipford & Yandle, 1997, cited by Mantzavinos *et al.*, 2004). Formal institutions, however, are imposed externally onto the community as the exogenous product of the evolution of relationships among rulers (Mantzavinos *et al.*, 2004).

According to Hatakka (2016, p. 17), the laws can sometimes be in conflict. Therefore, they do not give clear guidance for the actions in practice. It is

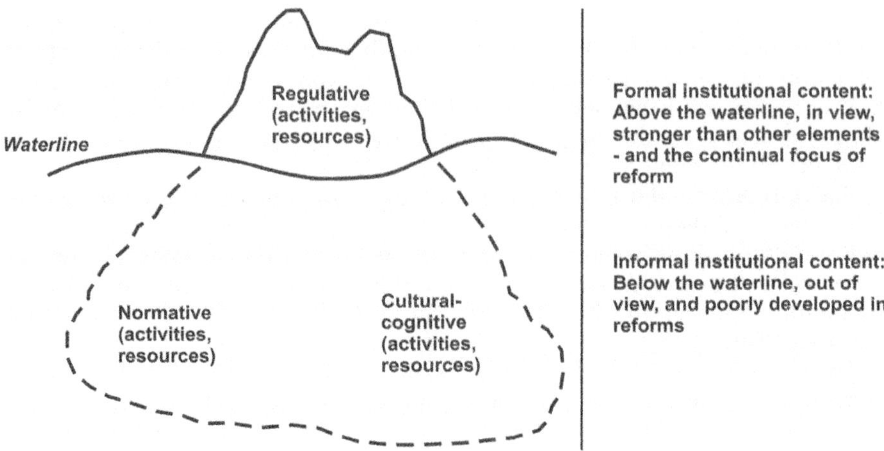

Figure 4.2 Reforms foster new institutions as a foundationless iceberg – the sinking kind. *Source:* Andrews (2013, p. 82). Reproduced with permission of the Licensor through PLSclear.

obvious that the formal rules require interpretation and conflict resolution. The incentives and consequences require planning and can lead to unpredictable outcomes. Hence, the effective mechanisms for ensuring compliance with the rules and regulations, and in addition, their correction procedures are necessary. Therefore, North (1993, p. 7) underlined that in addition to formal and informal institutions, the enforcement characteristics of both have also to be considered and included. The predominant rules in society are functioning correctly only when they are enforced comprehensively, and their regulatory compliance is monitored, promoted and ensured effectively.

Quite different institutional frameworks can be noted among indigenous nations. For instance, in southern Ethiopia, the traditional Borana and Konso communities have managed their wells and ponds for more than five centuries (Behailu *et al.*, 2016). It is probable that other respective communities have found their own ways of managing their water.

After this introduction to the rules of water services, the authors explore major firmly held myths of an institutional nature (Myths 13 to 17).

Myth 13: Water services have far too many stakeholders

Figure 4.3 It is difficult to create teamwork without a consensus.
Illustration: Pertti Väyrynen.

It is sometimes argued that the administration of water services is too fragmented, with too many stakeholders, and that it should be more efficient and centralised. While this may be true to some extent, the real issue might lie more in how to foster and sustain effective collaboration and the sharing of responsibilities, rather than focusing solely on the number of stakeholders involved (Figure 4.3).

The argument about having too many players does not necessarily take into account the fundamental purpose of water services: to serve the public through safe potable water and to make sure that wastewaters are properly managed to safeguard the environment. This means that through a variety of legislation water services are connected to many development sectors of society at various levels of administration such as health, environment, urban planning and construction. In other words, there are many players.

REALITY: Since water services are connected to practically all community activities, rather than additional centralisation of governance, partners should promote smooth collaboration between the different parties.

The issue of balance between a central power and democratic decentralisation and the role of citizens was elaborated in Figure 3.2. What should the appropriate balance be between centralisation and decentralisation in each case?

As of 2024, the European Union has 28 member countries. The idea behind its creation emerged after two world wars, based on the belief that it is better to work together than to fight against one another (European Union, n.d.). In practice it has proved to be quite challenging to find an appropriate balance of harmonisation of policies while at the same time considering the local conditions of individual nations (e.g., nutrients, Myth 11). The subsidiarity principle, one of the cornerstones in the Maastrict Treaty, and the Charter of the European Union (Kenton, 2024) is important for water services (Myth 1) where local conditions largely determine the solutions required (Juuti *et al.*, 2007). According to Raunio and Saari (2006), politicians and civil servants may often give reasons for their solutions on the basis of EU requirements whereas flexibility in adopting requirements to local conditions and the subsidiarity principle are not necessarily remembered.

A general development trend in many parts of the world in recent decades has been toward decentralisation. In some connections (e.g., Zajda, 2006) decentralisation has been connected to privatisation (Myth 17) and the role of the state, which in water services is not necessarily the case. In some cases (e.g., Verwoerd, 2024) the widely held assumption that decentralisation reduces corruption has been challenged indicating the need for a more nuanced understanding of its effects on public service provision, and thus on local conditions.

While the motivation for decentralisation will often vary from state to state, the following two sets of objectives are the most prevalent: (1) to design efficient public service delivery based on the principle of subsidiarity: services that can be effectively provided by lower spheres of government; to distribute public power broadly so as to achieve more effective and responsive government; to broaden access to government services and economic resources; and to encourage greater public participation in government; and (2) to construct a government structure in which diverse groups can live together peacefully; and to allow stakeholders representing a minority or marginalised regions to identify their space in the system, thereby underpinning the stability of the state by persuading them to remain loyal (Böckenförde, 2011).

Hooghe and Marks (2002) considered that beyond the adamant agreement that efficient governance must be multi-level, there is no consensus about how multi-level governance should be structured. Yet, Heikkila (2004) argued that a public service industry, or a local public economy, should have three components. These are jurisdictions and organisations that provide goods or services, jurisdictions and organisations that produce or supply goods or services, and jurisdictions and organisations that legislate and administer rules governing provision and production (Oakerson, 1999; Ostrom and Ostrom, 1991, cited by Heikkila, 2004, p. 102).

The overall role of local governments varies around the world. According to the Strong Elected Local Governments Index 2023 (with a scale: from 0 to 1), the lowest scores are found in Arab countries and China, more moderate scores in many African countries and Russia, and the highest scores in OECD countries (Our World in Data, 2024). Where applicable, there should be a bottom-up push from local governments to address existing challenges in water management. Mahajan and Rajankar (2024) stated that India has a long tradition of water harvesting systems and community-based management of water commons, thus demonstrating the need to understand local dynamics and ground realities.

A study by Ladner *et al.* (2016) measuring local autonomy in 39 European countries, showed an overall increase of local autonomy while also a significant variation between the countries. In terms of local economic development policies in Africa, OECD/UN and ECA/AfDB (2022) made the following five suggestions: (i) coordinated public policy packages rather than isolated initiatives; (ii) identifying and utilising a city's competitive advantages; (iii) specialisation enabling cities to generate economies of scale and increase productivity; (iv) stimulating economic development based on existing economic activity rather than an entirely new one; and (v) universities and other higher education institutions being key actors for local economic development.

A difference in the position of local democracy between Europe and the USA was noted by Hendricks *et al.* (2011), since local voter democracy through referendums, initiatives, opinion polls, consumer surveys and so on are more developed in the USA than in Europe.

According to the United Cities and Local Governments (UCLG) Policy Paper (2016), the 21st century faces daunting economic challenges (e.g., rapid urbanisation, mass poverty and unemployment especially in the Global South, changing production systems in the countryside such as soil depletion, water shortages, climate change, and population growth). There are also looming dangers arising from current economic turmoil and social inequality. The UCLG paper cited above argued that tackling these issues are the public institutions on the front lines, that is local and regional governments.

As an example of a water services' institutional framework, the case of Finland, subject to European Union regulations and directives, is presented in Figure 4.4.

Local governments (i.e., municipalities) are in charge of the general development and provision of water services. Water utilities owned by the municipalities or cooperatives produce water services within their designated service areas, whereas individual properties are responsible for their own systems. Domestic water quality is controlled by municipal health authorities (Hukka & Seppälä, 2004; Katko, 2016, pp. 166–171).

In Finland, the Ministry of Agriculture and Forestry is the body primarily responsible for water services development at the national level. The ministries of the Environment, Social Affairs and Health, Employment and the Economy, and Foreign Affairs also have their specific roles subject to the regulations and directives of the European Union (Katko, 2016, pp. 166–171).

For regional administrations in Finland, in 2010 the Centres for Economic Development, Transport and the Environment were established. They are also involved in the development, promotion and control of water management and services, with their foci subject to local needs. Regional councils prepare long-term regional plans which can include reservations for inter alia, major pipelines, treatment plants, and groundwater and artificial recharge areas (Katko, 2016, pp. 166–171).

The Finnish Environment Institute (SYKE) is an expert organisation that supports water pollution control and water management through research. It also collects data on water services and other water management issues and develops related tools as well as monitoring and assessing water resource fluctuations and the state of surface and groundwaters. Major water

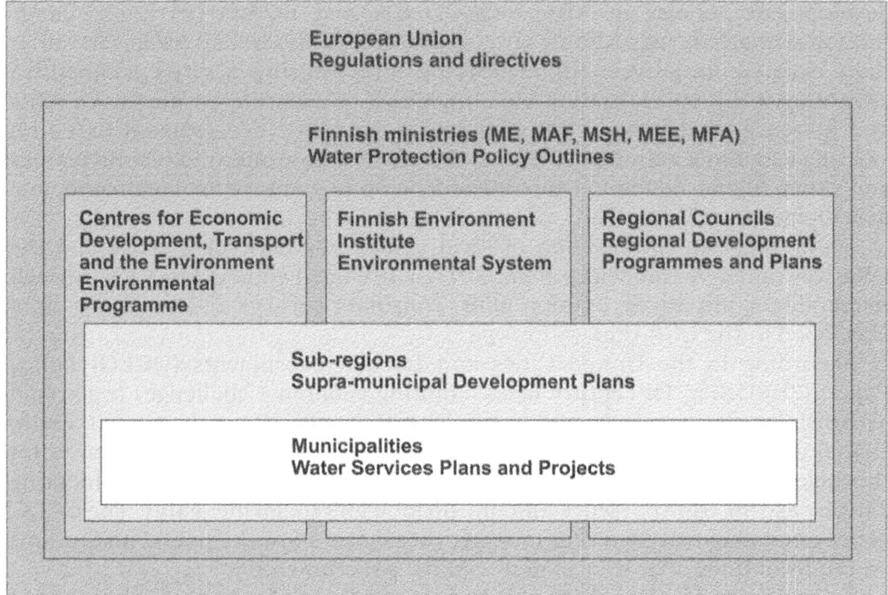

ME=Ministry of the Environment, MAF=Ministry of Agriculture and Forestry, MSH=Ministry of Social Affairs and Health, MEE=Ministry of Employment and the Economy, MFA=Ministry of Foreign Affairs

Figure 4.4 Overall institutional framework of water services in Finland in 2014. *Source:* Katko (2016, p. 168).

legislation can be classified into the three categories of health protection, built environment, and water resources use, protection and security (Katko, 2016, pp. 166–171).

Katajamäki (2011) pointed out that historically, Finland has been a locally focused civil society. While the role of civil society is emphasised, its actions are restricted by unbalanced competition requirements. According to Manninen, municipalities have had two options for development: mergers or municipal collaboration (Hynynen, 2011). However, as Ryynänen (2012, p. 273) pointed out, 'strictly concentrating on the size of a municipality will not solve the social problems of an ageing population'. Furthermore, the number of local governments in European countries varies a lot, with the highest number in France and Germany while the current Finnish figure of 309 municipalities in 2024 is relatively low bearing in mind the low population density at 5.5 million (Finland Population, n.d.).

In Europe, Newman and Thornley (1996, p. 30) determined that out of five major 'families', in terms of their legal and administrative traditions, out of the British, Napoleonic, Germanic, Nordic, and East European, the British (English) legal tradition is quite unique while that of Scotland is closer to that of other parts of Europe. This may explain why water services in Scotland were not privatised in 1989 as in England and Wales (Myth 17).

The diversity of local government role in Europe can be seen in the ways that water services are provided and produced (Figure 4.5). In Finland, municipalities by law provide or arrange water services. These services are produced by municipal utilities or consumer managed cooperatives within their mandated operational areas. Services and goods are often bought from the private sector, and in urban areas, water and wastewater systems are mostly integrated under a single entity (Myth 4, Katko *et al.*, 2010).

The setup in Lithuania resembles that of Finland and its private sector is also developing. In England, the water service systems are owned by private entities, whereby citizens are represented through regulatory systems, although municipalities no longer play a role. In France, municipalities own the systems, while most of the services are produced by private sector companies, although this has recently been challenged (Myth 17). It remains to be seen whether remunicipalisation will also take place in several larger cities of Central and Eastern Europe, which introduced private concessions or long-term operational contracts during the transition period (Katko, 2016, pp. 220–221).

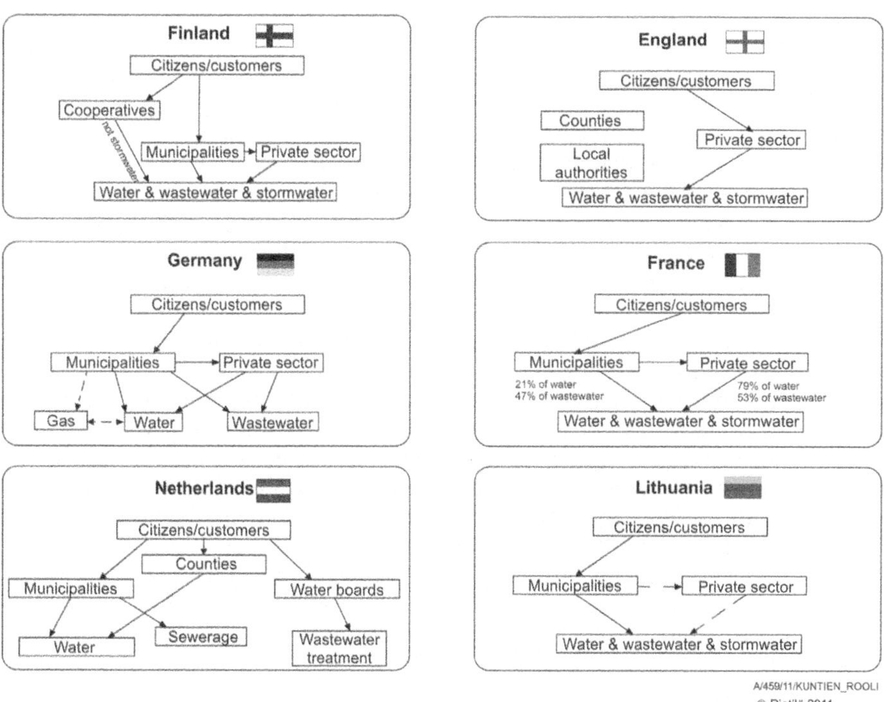

Figure 4.5 Institutional arrangements of water services and utilities in six European countries: the relationships between: citizens/customers; local and regional bodies; municipalities; the private sector; water and wastewater utilities; and cooperatives. *Source:* Pietilä (2013, p. 102); modified and cited by Katko *et al.* (2015, p. 304).

For historical reasons, in Germany, water and gas distribution are often assigned to a single municipal company (stadtwerke), whereas sewerage and wastewater treatment are more often managed by a municipal department. In the Netherlands, publicly owned water companies produce water supply and distribution, municipalities provide sewerage and drainage, and wastewater treatment is under the control of water boards (Pietilä, 2013, p. 102).

In Finland, the strategic functions of municipalities are decided mainly by municipal decision makers and can include: owner policy and control; target setting; strategic asset management; and water services development planning. The strategic functions of a utility include planning to carry out the strategy; business operations; development and investments; asset management, so on. The operational functions of the municipality are carried out mainly by municipal officials. These actions can include water services development, and decisions on service areas. Elected political decision makers or utility boards decide on water tariffs. Accordingly, the operational functions of a water utility are mainly carried out by its staff, consisting of core and non-core operations. Core operations are those performed by the utility whereas non-core operations can be, at least partly, outsourced. These roles and functions may vary (Katko, 2016, p. 205).

Currently in Finland the trend is to make municipality owned water utilities more autonomous through corporatization, which water utility experts seem to favour. However, there are potential concerns since corporatisation may reduce transparency and openness and perhaps the chances of integrating urban services management, which is important at least for stormwater management. According to Marin *et al.* (2010) in Burkina Faso a well performing public water utility was achieved that ensured the autonomy and accountability of water producers as well as viable tariff and investment policies. In Italy, Landriani *et al.* (2019) found that in the decorporatisation of a municipal water utility, it is possible to find an equilibrium between economic and social interests.

Another area of potential development is better involvement of citizens and residents. In a socio-hydrology study by Haeffner *et al.* (2018), residents showed greatest concern about future water shortages and high water costs whereas their leaders were concerned most about deteriorating local water infrastructure. Transition to sustainable water governance could include water user involvement in local water systems. One way of promoting this by water utilities is to organise water user advisory boards without giving them decision-making power as exist in the USA (Becker, 1993) and occasionally also in Europe.

Corporatisation of public utilities is one option for developing water services as an alternative to privatisation (Myth 17). Some public utilities claim to operate on a 'public' basis in their home country while pursuing for-profit private contracts elsewhere, thus employing 'double standards' (McDonald, 2022, p. 43). In Finland, the corporatisation of municipal water utilities has been a growing trend. While utilities view increased autonomy as a positive development, it raises questions about how to ensure adequate collaboration between water services and other municipal actors, as well as public access to relevant information. For two decades, the authors, (among others), have questioned why municipal utilities are allowed to practice hidden taxation while lacking funds for necessary repair and renewal investments (Hukka & Katko, 2007; Vinnari *et al.*, 2005).

According to McDonald (2016, p. 108), corporatised entities make up a significant portion of the public sphere in many Western European countries and are widespread in North America and a dominant organisational trend in Asia, Africa, and Latin America. Furthermore, (McDonald 2016, p. 107) noted that the recent resurgence of corporatisation has been heavily influenced by neoliberal theory and practice. The literature on corporatisation is highly polarised, as demonstrated by Marin *et al.* (2010) in Burkina Faso and Landriani *et al.* (2019) in Italy.

McDonald (2016, p. 108) advocated for a middle-ground approach that considers concerns about the commercialisation of public services while acknowledging the potential for corporatisation in more progressive, equity-oriented ways.

To create mechanisms for active user consultation and to ensure that stakeholders are part of the decision-making process, Water Utilities Advisory Boards are quite commonly used in the USA (Ozekin, 2024). These provide input to the city council or utility board, offering advice or recommendations without decision-making power. There is evidence about such bodies from Finland and Germany (Bauby *et al.*, 2018, p. 7), although they have much wider potential for utilisation in other countries.

It is obvious that in water services – along with acceptance of biological and cultural diversity – institutional diversity (shortened to 'insdiversity' by Katko *et al.*, 2022a, p. 604), is also important as argued by Nobel Laureate Ostrom (2005). It is feasible to find out about the most applicable playgrounds (Figure 4.1) and institutional arrangements. In this way, we cannot avoid the large number of stakeholders in water services.

Myth 14: Water is a basic human right and therefore it should be free

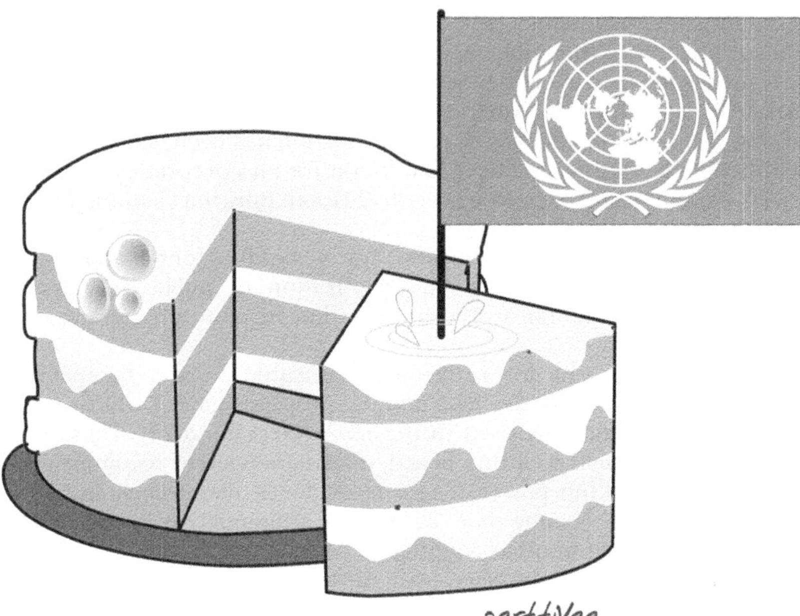

Figure 4.6 Water is a basic human right, although other pieces of the water services cake, first and foremost recovering the costs incurred, must be considered and introduced. *Illustration:* Pertti Väyrynen.

The United Nations World Water Development Report (UN, 2021) observed that unlike most other natural resources, the 'true' value of water is recognised as being extremely difficult to determine. Therefore, the overall importance of this vital resource is not appropriately reflected in political attention and investments. This not only leads to inequalities in access to water resources and water-related services, but also to inefficient and unsustainable use and degradation of water supplies themselves, which affects the fulfilment of almost all the SDGs, in addition to basic human rights.

Water management is, however, a social issue for all that transcends party politics; there is very little else that unites citizens in this way. Water has several characteristics: it is at the same time a fundamental right, an indispensable product, economic good, and a public 'commodity' (Figure 4.6). For this reason, water cannot be treated like any other commercial commodity. Instead of arguing about the price of water, much more should be said about its value and importance for the wellbeing of communities and citizens. An international comparative study found that Finns consider water, in addition to forests, to be Finland's most valuable natural resource and a resource of the national economy (Kemira, 2020).

It can be justifiably argued that the right to water and sanitation is *an implicit human right*. Since people cannot survive without water, the right to water can be seen either as a universal ethical and moral right or as a fundamental precondition for the enjoyment of several other explicit human rights, such as the rights to life, an adequate standard of living and health, education, housing, work, and protection against cruel, inhuman and degrading treatment or punishment. Therefore, the human right to water cannot be considered to only cover drinking water; the most crucial aspects are related to those water functions and uses that are essential for human survival and dignity such as adequate sanitation services or the preservation of healthy aquatic ecosystems in addition to access to safe water supply (Hukka *et al.*, 2010a, p. 241).

In a study by Rajala *et al.* (2019), international water and environmental students defined the importance of water using six predetermined criteria as follows: water is: a basic human right (31%), a natural resource (25%), an economic good (15%), a public and social good (both 11%), and a cultural resource (7%) (Figure 4.7, n = 241). Responses from the different countries were surprisingly similar, although the countries differ greatly in their circumstances and levels of development. The result shows a fairly general order of importance despite the fact that the implementation of water services depends crucially on

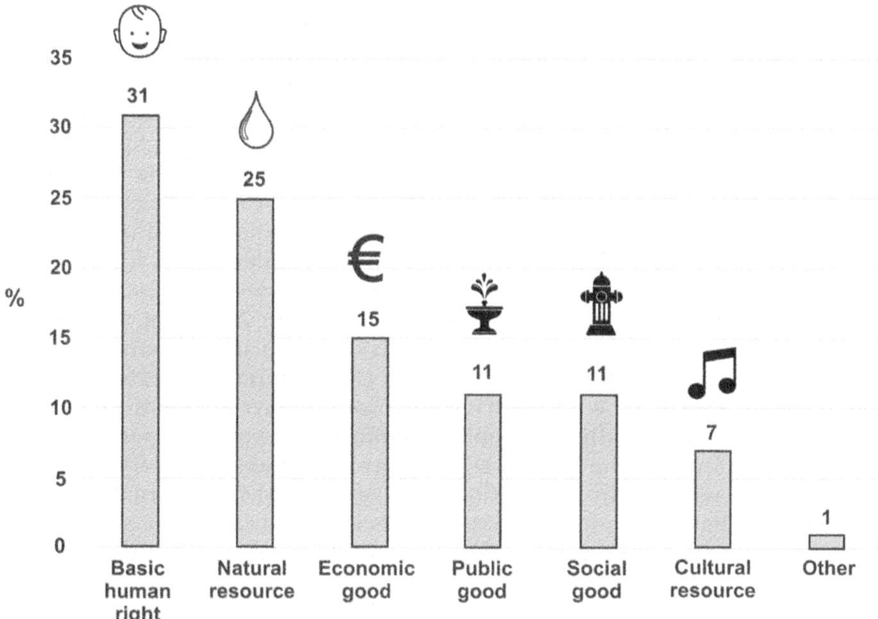

Figure 4.7 Overall results from seven countries and one region, compiling global ranking on perceived priorities related to water.
Source: Rajala *et al.* (2019, p. 12, modified), illustration: T. Katko 2024.

local conditions. Water is, therefore, simultaneously a right, a natural resource and a multipurpose commodity, as also observed for example by Peda and Vinnari (2020, p. 17).

Gradual recognition of water as a basic human right

The first important step toward broader recognition of water rights was taken in the UN Water Conference in Mar del Plata, Argentina in 1977, when for the first time, governments recognised water as a right, declaring that 'All peoples, whatever their stage of development and social and economic conditions, have the right to have access to drinking water in quantities and of a quality equal to their basic needs' (Benöhr, 2023, p. 57; UNW-DPAC, 2011, p. 1). In 2002, the UN Committee on Economic, Social and Cultural Rights adopted General Comment No. 15 on the right to water. The Committee stressed that water is a social and cultural right and not just an economic commodity. The human right to safe drinking water was first recognised by the UN General Assembly in 2010, when 122 countries formally acknowledged this right (UNW-DPAC and WSSCC, 2011).

The United Nations Human Rights Council (HRC) adopted a consensus – building on the resolution by the United Nations General Assembly – in September 2010 and formally acknowledged the importance of safe drinking water and adequate sanitation to societal development by affirming that access to water and sanitation is a human right. The HRC's resolution states that 'the human right to safe drinking water and sanitation is derived from the right to an adequate standard of living and inextricably related to the right to the highest attainable standard of physical and mental health, as well as the right to life and human dignity'. This resolution confirms that the rights to water and sanitation are part of existing international law, and are legally binding on states. The resolution, however, fails to provide guidance on how progress is sufficiently monitored, or how the capital, and operations and maintenance (O&M) costs of water and sanitation services infrastructure are provided while maintaining affordable prices for the poor (WWAP, 2012, p. 35).

The UN General Assembly explicitly recognised in 2015 that the human right to sanitation is a distinct right. In terms of the normative content of the two rights, the HRC reaffirmed in 2018 that: (1) The human right to drinking water entitles everyone, without discrimination, to have access to sufficient, safe, acceptable, physically accessible and affordable water for personal and domestic use; and (2) The human right to sanitation entitles everyone in all spheres of life, without discrimination, to have physical and affordable access to sanitation that is safe, hygienic, secure, socially and culturally acceptable, and provides privacy and ensures dignity (UNECE and WHO, 2019).

Access to water includes the right to abstract water and receive adequate water services. The obligations of states concerning the right to water would, however, be regarded as goals, not as specific entitlements (McCaffrey, 1992, cited by Hukka *et al.*, 2010b, p. 241). Article 17 of the Berlin Rules does not stipulate every detail of what a right of access to water entails, but it does provide the basic principles within that right. In accordance with Article 17 'every individual has a right of access to sufficient, safe, acceptable, physically accessible, and affordable water to meet that individual's vital human needs' (ILA, 2004, p. 22).

Ensuring access to both water and sanitation as human rights means that access to safe water and basic sanitation is a legal entitlement, rather than a commodity or service provided on a charitable basis. Furthermore, universal access to sanitation is 'not only fundamental for human dignity and privacy but is one of the principal mechanisms for protecting the quality' of water resources (UNW-DPAC and WSSCC, 2011). For example, in Finland, which is a considerably wealthy nation, citizens outside the service areas of water undertakings – in conformity with the Water Services Act (681/2014) – enjoy only the right to abstract water, not the right to water services provided by their respective municipalities (Hukka *et al.*, 2010a, p. 247). When a property is located outside the utility's designated service area, the property's obligations to take care of water supply and wastewater treatment and disposal are not stipulated in the Water Services Act (WSA) but in other legislation linked to issues such as health, environment, land use and building, and water resources (Beilinskij, 2015, p. 10).

The fourth guiding principle of the Dublin Statement on Water and Sustainable Development (ICWE, 1992) stated: Water has an economic value in all its competing uses and should be recognised as an economic good. The United Nations Conference on Environment and Development (UNCED) in Rio de Janeiro in 1992 recommended that states should introduce sound financial practices, achieved through better management of existing assets as well as to introduce water tariffs, considering the circumstances in each country (UNCED, 1992). Many critics have argued and highly contested this fourth Dublin Rio principle by saying that this is just an obsolete example of the neoliberal global policy. The full text of Principle 4, however, stipulates: 'Within this principle, it is vital to recognise first the basic right of all human beings to have access to clean water and sanitation at an affordable price.'

All in all, the human rights framework does not provide for the right to free water. Yet, in certain circumstances, access to safe drinking water and sanitation might have to be provided free of charge if the person or household is unable to pay for it. It is a state's core responsibility to ensure the satisfaction of at least the lowest essential levels of this right, and access to the minimum requisite quantity of water. Water and sanitation services need to be affordable for all and, therefore, states should adopt the necessary measures, including appropriate pricing policies such as free or low-cost water. People are expected to contribute financially or in kind to the extent that they are able do so (OHCHR *et al.*, 2010, pp. 11–12; UNW-DPAC and WSSCC, 2011).

Under Article 9 of the Water Framework Directive (WFD), EU Member States are obliged to apply the principle of recovering the costs of water services, including financial, environmental and resource costs, in line with the principles of the polluter pays and the user pays. As costs are, to an extent, passed onto customers, Member States are also required to set up water pricing mechanisms, which should provide adequate incentives for water users to use water resources efficiently. The WFD also mandates Member States to report on cost recovery and the enforcement of adequate incentive pricing (Farnault & Leflaive, 2024, p. 8; EurEau, 2017; Benöhr, 2023, p. 62).

In Finland, the utility operating under the WSA should be able to provide services cost effectively and adequately, and the services charges should be affordable, fair and equal (Beilinskij, 2015, p. 13). The rationale for charges and cost recovery must also be transparent. In the long term, water services and stormwater charges must cover all repair, renewal and greenfield investments as well as any operating costs incurred. In accordance with the cost recovery principle, the charges must correspond as far as possible to the actual costs of the services. Cost recovery ensures that the utility has economic and financial capability to fulfill its tasks through sufficient revenue generation. Therefore, only a reasonable rate of return may be included in the charges. In addition to the financial cost, environmental and resource costs may be included in accordance with Article 9 of the WFD. Notwithstanding this cost recovery principle, the WSA permits the financial support for water services from the funds of the municipality, the state and the EU (Beilinskij, 2015, pp. 34–35).

The facts have revealed that inadequate cost recovery is still one of the major obstacles towards sustainable drinking water supply and sanitation services in developing countries (Cardone & Fonseca, 2003). In this context, water pricing first and foremost is a financing mechanism: it generates revenues that can be used to maintain, renew and extend the infrastructure, when and where appropriate (OECD, 2010). Although the full cost must be estimated and benefited for rational allocation and management decisions, they do not necessarily need to be recovered entirely from the users. The cost, however, will have to be borne by someone. Estimation of full cost may be very difficult especially in situations involving conflict over water, where attempts should be made to at least provide an estimate as the basis for allocation (GWP, 2000).

REALITY: Water as a natural resource is free, but regulatory governance, provision and production of water services and the protection of the quality and availability of water resources inevitably cost money. This needs to be covered keeping in mind the requirements of water and sanitation as a human right.

To ensure affordability of water services – one of the dimensions of the human right to water – the European Federation of National Associations of Water and Wastewater Companies (EurEau) recommends that customers who cannot pay their water bills should be supported using social policy instruments and not to be granted an exception regarding water bills (Rasenberg, 2021). In Finland, to ensure the human right to water and sanitation, the water utility can only discontinue supplying water to its customer or conveying waste and stormwater from the premises, if that customer has neglected to pay for services or has breached their obligations in some other way. The discontinuation of water services can be done carried out five weeks from giving notice to the customer. In practice, the utilities generally come to an agreement about payment of bills in arrears, and only cut off the water supply when the loss of revenue has accumulated considerably over a long period of time. However, force majeure – for example, a serious illness or unemployment and where the customer has informed the utility about their difficulties to meet payments – should always be taken into consideration (Beilinskij, 2015, p. 32).

Seppälä and Katko (2003, p. 235) argued, however, that full cost recovery (FCR) should be a key policy objective whenever pertinent water infrastructure investments are considered. Water pricing must be *regulated*, however, to safeguard full cost recovery and to ensure that citizens and businesses get affordable, fair and equal prices (Hukka & Katko, 2021, p. 70). Keeping water tariffs artificially low for all customers may not ensure the affordability of water services to low-income households and may result in a vicious circle of underfunded service producers, insufficient investments, ageing and collapsing infrastructure and deteriorating services. In this case, the quality of water services would deteriorate and future users would not be able to enjoy the same quality at a similar level of affordability, which further reduces the perceived benefits and thus also the users' willingness to pay. Collapsing water systems can impact lower-income users the most, and especially those who currently do not have access to water services, as low tariffs (in the absence of reliable transfers that can make up for missing revenues) may prevent the extensions of networks to poorer communities (EurEau, 2016; OECD, 2009).

Hukka and Katko (2015, p. 44) presented a simplified vicious circle of an unviable water services pricing regime (Figure 4.8), where ignorance is the most likely issue that should be tackled.

Among other places, in the East African region, many new independent governments had ambitious plans and policies in the 1960s including the free water policy at least in rural areas. In Tanzania, following independence in 1961, the Government started to finance all water supply investments in 1965, and in 1970 extended this to cover operations and maintenance (O&M) costs in rural areas. Urban users were in principle supposed to pay for their water, although the tariffs were not adjusted until 1988. Warnings by several experts on the unviability of this policy were ignored. The Nordic bilateral agencies in particular supported regional rural water supply projects until the early 1990s but could not influence the free water policy. This policy relating to rural areas was formally abandoned in 1991 in favour of cost-sharing (Mashauri & Katko, 1993).

In Kenya, tariffs declined almost continuously until the mid-1990s whereafter they started to be gradually adjusted (Seppälä & Katko, 2003). However, the level of O&M costs coverage ratio of a regulated utility should be over 150% to meet full cost recovery. According to the Water Services Regulatory Board (WASREB), the O&M costs coverage ratio decreased from 96% in 2021/22 to 95% in 2022/23. This indicator of an overall decline for the third consecutive time demonstrated the increasing difficulty experienced by Kenyan regulated utilities to fully cover their O&M costs, especially for the small- and medium-sized categories (WASREB, 2024).

OECD (2020 cited by Kuulas *et al.*, 2020, p. 65) produced estimates of the required water services investment by all EU Member States by 2030 in order to comply with the EU directives that bind the water utilities. This study reveals

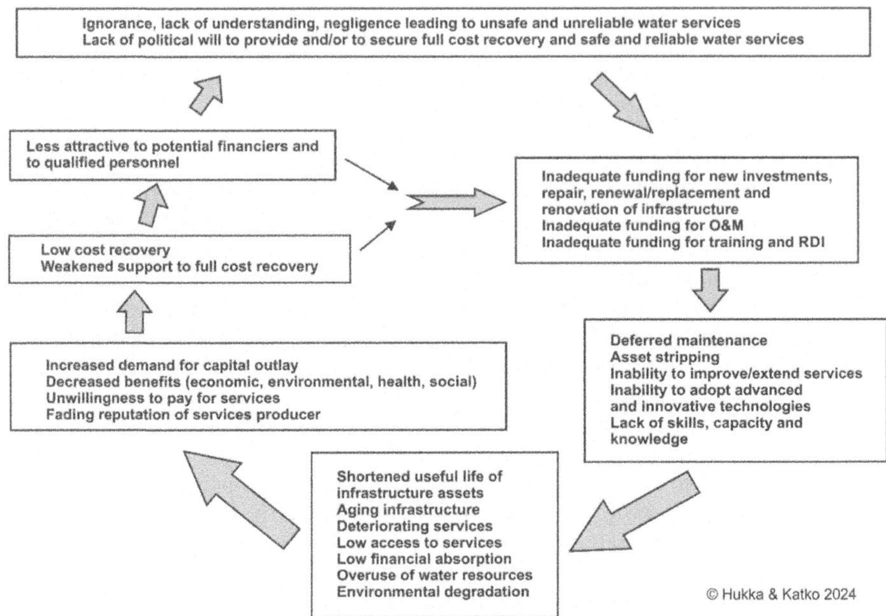

Figure 4.8 Vicious circle of an unviable water services pricing regime.
Source: Hukka and Katko (2015, p. 44), modified in 2024.

that all countries need to increase their current annual investments by more than 20% in order to achieve compliance. Some countries face the challenge of much larger financing needs, such as Romania (+180%), Bulgaria (+100%) and Finland (+85%). These estimates on future financing needs and capacities reveal increasing challenges for cost recovery (Farnault & Leflaive, 2024, p. 25). In Finland, it was estimated that the total annual investment needed will be approximately EUR 777 million by 2040. Consequently, if water and wastewater tariffs were increased to meet this, they would need to be increased on average by approximately 50% (Kuulas *et al.*, 2020, p. 67). In any case, the costs must be covered to safeguard the sustainability of water services and the resilience of the water utility.

The Global Commission on the Economics of Water (2023) proposed a basic market approach to tackle global water challenges, introducing the notion of water as a global common good in a report launched at the UN Global Conference. They argued a central role for private financing in the provision of water supply and sanitation services. As shown in Myth 17, this has not succeeded despite marketing it for decades. Heller *et al.* (2023) called for international cooperation and solidarity to address power asymmetry, inequalities, and unaffordable access, and to recognise local conditions, the key role of public finance, and the human rights at the core of the water agenda.

Myth 15: All water service problems will simply be solved by developing technologies

Figure 4.9 Any problems can be solved merely by good technical planning and modelling. Or can they?
Illustration: Pertti Väyrynen.

The conventional and quite understandable way of perceiving technology is to focus on artefacts and equipment, pointing out good technical planning and modelling (Figure 4.9). However, experiences from various countries and corners of the world show that major bottlenecks in water services are often connected with the application of technologies, financing, operations and related institutions, that is the rules of the game (Figure 4.1).

The word 'technology' is often used and applied to various aspects of our daily lives. But what is technology and how should we understand it? Leppälä (1998) offered a wider definition to cover (i) technological artefacts, (ii) procedures, and (iii) the knowledge required to apply both of these. If technology is understood merely as technological artefacts and equipment in its use, potential major problems may arise.

In addition, technology should include institutions as discussed in the introduction of chapter four. This broader approach to technology is highly appropriate as regards water and water services, as explained below.

In the late 1970s, before the United Nations International Drinking Water Supply and Sanitation Decade (IDWSSD) 1981–1990, Pacey (1977) explored the dimensions of appropriate technology, and concluded that technology alone is not enough and various criteria for technical, social and economic appropriateness are also needed. Pacey (1983, p. 49) introduced two major spheres for examining technologies: the user sphere and the expert sphere and argued that 'good technology' should make use of both. In addition to technical aspects, it is important to consider cultural and organisational dimensions. Furthermore, in accordance with the Helsinki Term Bank for Arts and Science (2024), appropriate technology is defined as locally manageable technology.

In technological development, the theory of path dependence is often used, referring to the influence of past decisions on future societal processes and outcomes, shaping adaptation strategies and hindering the implementation of alternative solutions. This concept created by economists was later applied in other areas. Melosi (2005 cited by Melosi, 2008, p. 8) suggested that decisions made about sanitary systems in the nineteenth century had fundamental impacts on cities more than a hundred years ago, such as combined sewers (Myth 4) or raw water selection (Myth 3). Joosse and Teisman (2021) argued, however, that path dependency shows up as a weakness of simplification, while path-creation is seen as a strength of complexification.

Path dependence is often explained through technological logins, which are often viewed in a negative light as having limited options or leading to a too fixed path. Yet, such decisions can be either negative or positive depending on whether they prove to be clever in the long run or not. One positive example of such a positive path – not to use lead pipes in Finland – was mentioned in Myth 10.

Future researchers, or futurists as they were called earlier, have pointed out the evolutionary nature of development. Accordingly, development and technology are not deterministic and bifurcations or turning points are bound to occur. In terms of water, bifurcation also refers to upstream water bodies of watercourses which may flow to two different catchment areas, as illustrated in Figure 4.10. Sometimes *revolving paths* may also occur as in the selection of urban raw water sources in Finland.

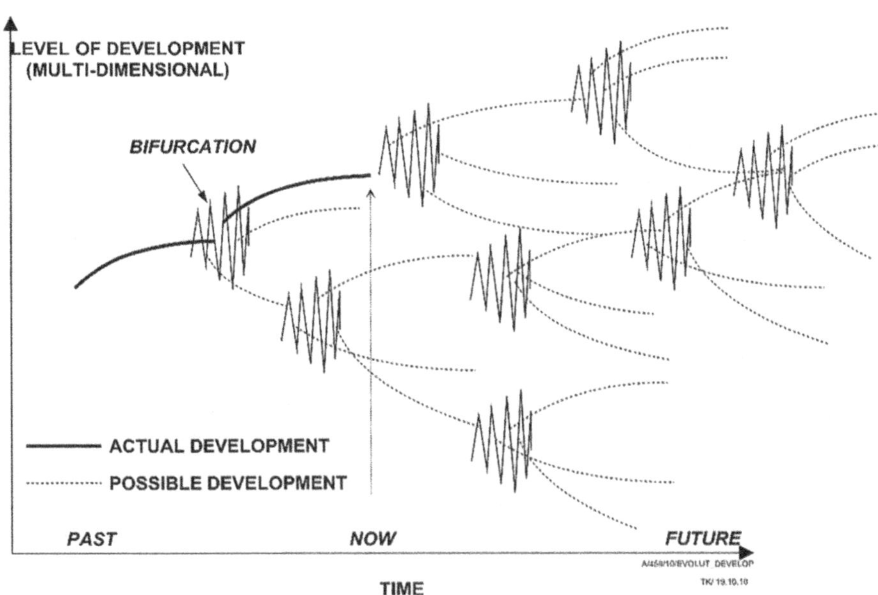

Figure 4.10 Bifurcations in technological development, sometimes involving revolving paths.
Source: Mannermaa (1991, p. 240), modified by Antila *et al.* (2013, p. 20).

Perhaps the most straightforward theory is that technical systems will expand continuously, with hardly any limitations. This idea called Large Technological Systems (LTS) (Hughes, 1987) is to a large extent applicable to road and electricity networks but not to more local services such as water.

American historian and Professor Melvin Kranzberg (1917–1995) defined six laws of technology as follows (annotated by the authors):

(i)   'Technology is *neither good nor bad*; nor is it neutral.'
(ii) Invention is the mother of necessity. 'Every technical innovation seems to require *additional technical advances* in order to make it fully effective.'
(iii) Technology comes in packages, *big and small*. 'The fact is that today's complex mechanisms usually involve several processes and components.'
(iv) Although technology might be a prime element in many public issues, *nontechnical factors* take precedence in technology-policy decisions.
(v)   All history is relevant, but the *history of technology* is the most relevant. 'Although historians might write loftily of the importance of historical understanding by civilized people and citizens, many of today's students simply do not see the relevance of history to the present or to their future.'
(vi) Technology is a *very human activity* and so is the history of technology. 'The function of the technology is its use by human beings – and sometimes, alas, its abuse and misuse' (Kranzberg, 1986).

In terms of water services development, 'the Kranzberg laws' seem to be largely relevant. Probably the biggest 'technology leap' in Finnish WSS evolution was the shift from wooden to plastic pipes when domestic manufacture of polyethylene pipes began in 1954. At first, this occurred on a small scale in rural water supply and gradually spread to a larger scale and urban water, stormwater and wastewater systems. This development led to the creation of one of the leading plastic pipe manufacturing companies in Europe (Katko, 2016, p. 213).

The Social Construction of Technology (SCOT) theory tries to understand the links between social and technical processes by seeing them as human or social constructions. Technology is shaped by human engineers, market forces, consumer needs and demands, individuals and groups as well as power politics and the internal structure of technology (Bijker *et al.*, 1987). Mattila (2005, pp. 31, 60, 83), for example, applied this to the management of on-site sanitation.

Good planning and modelling can help, assuming that adequate perspectives are taken into account. For sustainable development it is important to consider the three core elements of economic growth, social inclusion and environmental protection in a balanced way. These elements are interconnected and crucial for the wellbeing of individuals and societies. For instance, marine waters should not deliberately be polluted by granting permits of exception to mining companies for the special minerals needed for manufacturing rechargeable batteries for electric cars. When applicable, nature-based solutions such as managed aquifer recharge (MAR, Myth 3) should be used.

> **REALITY**: Water services provision requires distinct and abundant technology. Water services as a whole form a multifaceted sector, the problems of which can be viewed from several angles. Technology is to be understood in its wider context and is in addition to the technological equipment required to cover procedures and knowledge to create both.

According to UNEP (2012, p. 15) the transition to a green economy presents a triple challenge (Figure 4.11). First, there is the focus on the economy, finding ways to increase prosperity without increasing resource use and environmental impacts, that is being more resource efficient. However, this is not sufficient and the second challenge is the need to maintain ecosystem resilience. The third challenge is human wellbeing, including health, employment, job satisfaction, social capital and equity. This also includes a fair distribution of the benefits and costs of the transition (UNEP, 2012, p. 15).

In balancing environmental, economic and social challenges, the green economy concept evidently has much in common with various models of sustainable development, which recognises the triple challenge of economic efficiency, ecological sustainability and social equity (UNEP, 2012, p. 15). To create wellbeing for the wider society in terms of water services, Heino (2016, IV) suggested that there needs to be a paradigm shift to consider more the changing and ever more complex operating environment. To achieve this, the ways water services are perceived and understood need to change from production orientation to a more service-oriented paradigm.

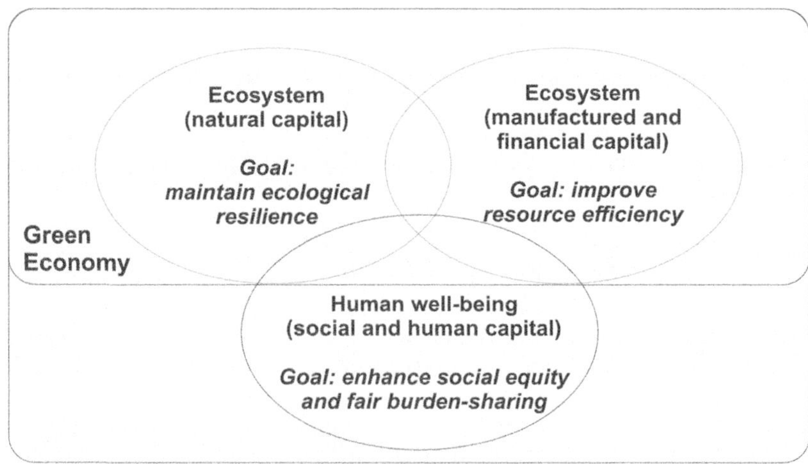

Figure 4.11 Creating a green economy: a triple challenge.
Source: UNEP (2012, p. 15). Permit granted by UNEP, IRP Secretariat.

Knowledge management has become one of the key concerns of organisations in recent decades. For water utilities knowledge management is also a critical factor. Sandelin *et al.* (2019) stated that as a strategic asset, knowledge should be managed by water and wastewater utilities since further development will need crucial understanding of previous procedures and practices.

The need to understand water services development and related technologies from a wide perspective has obvious implications for water services education and research. Hukka *et al.* (2007) argued that the bias in favour of a positivistic approach and natural sciences in water research prevents finding adequate answers and solutions to wider water governance challenges and institutional and management issues. Water research should, therefore, be expanded to include diverse pluri-, cross-, inter- and multidisciplinary approaches as joint efforts, while individuals could be encouraged to seek transdisciplinarity.

Water services are most naturally included in combined civil and environmental engineering programmes. Furthermore, the SDGs demonstrate that water services (Figure 1.1) have connections to many disciplines such as hydrogeology, hydrology, urban planning, geography, history, political science, sociology, sustainability sciences and futures research.

Marjoram (2010, p. 3) estimated that the shortage of engineers in developing economies hampers development, especially for the poorest, who still lack adequate water and sanitation services. For example, to meet the Millennium Development Goal (MDG) target for safe drinking water and basic sanitation by 2015 would have required 2.5 million new engineers and technicians in sub-Saharan Africa alone.

The UN World Water Development Report 2016 (UN, 2016) estimated that 78% of jobs constituting the global workforce are dependent on water; 42% are heavily and 36% moderately water dependent. According to Joshua Newton, the skills that will continue to grow in the water sector include: technical skills, negotiation and conflict resolution, communications and writing, programme management, economics and finance, multiple languages for international work, and language to translate within disciplines in a multidisciplinary world (WaterFront, 2022). Grigg (1996, p. 4) contended that to solve the problems of water industries, each nation 'will need effective water industry managers who straddle with the technical, administrative and political worlds'.

In spite of the obvious challenges, water services education and research still focus strongly on water and wastewater treatment, which are valuable for community health and environment. Although 80% of the value of water utilities' assets are due to the networks (Myth 5), probably 90% of the research resources, however, go to treatment-related issues (Seppälä, 2019). This contradiction should certainly be corrected. It is likely that most research on water supply and sanitation follows quantitative and positivistic traditions while more qualitative and problem-oriented themes on water services management, institutions, policy and governance are emerging (Hukka *et al.*, 2007). However, all disciplines have their own traditions whereas many sector challenges require multi- and interdisciplinary approaches. Therefore, all research should be assessed against its own epistemological and ontological approaches (Katko, 2016, p. 188).

Regarding continuing education, the authors have been involved in developing and running a Water Services Leadership and Development (WASLED) programme at Tampere University, Finland, in cooperation with representatives of major national water services. Since 2009, this ten-month programme has been run seven times for mid-career water services experts with a total of 123 participants so far. The feedback includes increased networking and interaction, new perspectives, and more holistic views on water services (Inha *et al.*, 2019b). As current education and science policy strongly favours publishing mainly peer reviewed papers at the cost of other activities, it is increasingly important to promote continuing education on Management, Institutions, Policy and Governance Affairs (MIPGA). These fields should also be covered by undergraduate, post graduate and doctoral programmes to offer at least a basic knowledge of their importance and key principles.

One option to find a reasonable balance is to conduct 'pracademic research' (Price, 2001), which would provide answers to questions arising from society through scientific research.

Myth 16: Major water service constraints can be solved simply by merging utilities

Figure 4.12 Merging water and energy is not that simple as their special features need to be taken into account.
Illustration: Pertti Väyrynen.

It has been argued that the constraints of water services are caused by the large number of water service producers (municipal utilities or community managed water cooperative utilities). Accordingly, it has been proposed that these problems can be solved merely by merging utilities and making them larger. To achieve larger organisations, energy and water utilities have sometimes been merged (Figure 4.12).

In the Netherlands in 2004, the Government banned private sector provision of water supply, although in practice, drinking water companies contract out many services (e.g., customer relations and repairs) to the private sector (OECD, 2014). In water supply, many mergers of utilities have occurred although there have been far fewer in wastewater services. Merging of water supply utilities has been affected by the privatisation of energy utilities, although this was hindered by legislation (Kuivamäki *et al.*, 2023).

Public policy makers have raised such arguments in Finland over the last two decades (Katko, 2016, pp. 136–137) and in Kenya, in particular after the amendment of the Water Act (RoK, 2016, Section 97; WASREB, 2018). The assumed benefits of economies-of-scale are most likely to be taken for granted and, therefore, there is most no clear need for any further analysis and discussion. However, it is questionable whether the strengths and opportunities of these mergers are that simple and straightforwardly achieved? It seems that this belief or ideology of the unlimited benefits of economies-of-scale – without empirical evidence – has unfortunately spread to other sectors in society. There is an additional concern that small water utilities in particular may not have adequate know-how and resources to meet the increasing requirements.

In the early 2000s, Finland experienced a boom of municipal merging. In 2007–2012, approximately 60 municipal mergers were undertaken, reducing the number of municipalities by more than 100 down to 320 (Mäntysalo *et al.*, 2017). There have been further mergers since, reducing the number to 309 in 2021–24 (Kuntaliitto, 2024). Notwithstanding, the merger of two poor municipalities does not automatically produce a viable one as was noted by Manninen, a former minister (Hynynen, 2011).

According to Vakkuri (2010) the discussion on the economies of scale regarding municipalities raises many common truths argued on the basis of scientific theory, while forgetting the ambiguity and partial contradictions of viewpoints that occur in scientific discussion. In the USA, an opposite phenomenon to urban growth has been noted; since 2011 fewer people have been moving out of rural areas and more people have been moving in (Cromartie & Vilorio, 2019; Gavin, 2023).

Vakkuri (interviewed by Laurinolli, 2011) warned that local governments are overly concerned with municipal finances and criticised the blind belief in economies of scale. Productivity and cost-effectiveness do not depend on population size, but on how services are *organised*. Municipalities that are too large or too small seem to be ineffective which means there is no correct number of municipalities.

In 2010, new legislation relating to universities (24.7.2009/558) was enacted in Finland, promoted by the idea of ensuring autonomy to universities. According to Piironen (2013), a conceptual shift took place at the turn of the millennium based on managerial values and top-down organisation that emphasised the potential of universities to operate on the education market as would any enterprise. Thereafter merging of universities took place in many locations based on the argument of making them more efficient. In reality, universities have become strongly top-down organisations with increased bureaucracy, centralised leadership and loss of autonomy (Nevala *et al.*, 2023).

In 2017, Finnish assessment of the state of infrastructure (ROTI) recommended that water services needed a structural change 'to bring together the regionally dispersed functions which belong to the same set of services to create larger administrative units' without taking into account evidence or looking at other options (ROTI, 2017). This was repeated in a study commissioned by the Prime Minister's Office (Berninger *et al.*, 2018). Later a process of water services development reform appeared on the agenda but even here, the argument in favour of structural change has occasionally surfaced.

To our knowledge, in the legislation on the Finnish renewal process of water services, these challenges have been approached by: (i) identifying and accepting the diversity and variety of water service sectors and conditions, (ii) accepting the challenges, their variation and order of magnitude, and (iii) tailor-made means and actions to be selected based on the two points above.

Although the development of water services has progressed towards larger utilities in Finland, no organisational form can solve the greatest challenges of water services. In particular, these include inadequate investment in renovation and renewal, the need to prepare for vulnerability and resilience, inadequate human resources and competence, and insufficient input in research and development (Katko, 2016, p. 137; Juuti *et al.*, 2019; Katko *et al.*, 2022a, p. 155). Since the 2001 Water Services Act (no. 2001/119) it has been possible for municipalities to make a reasonable rate of return, which is not, for some reason, specified and regulated. Thereby, the larger cities are able to practise hidden taxation while not putting adequate resources into renovation (Hukka *et al.*, 2010b).

Are large water utilities always functioning on a sustainable and viable basis? The answer is most probably no, because in England for example, some very large and regulated private water companies are facing government bailouts. The UK water industry has come under scrutiny, particularly concerning the challenges faced by company finances and performance. Alarming and underlying issues of huge debts, endless fines for environmental violations, and regulatory pressures have brought these concerns to the fore. External investment comes at a time when rising inflation and interest rates are making debt servicing increasingly burdensome for UK water companies (Bluechain Consulting, 2024).

Of the 16 largest water companies in England and Wales there are 10 companies categorised as requiring an increased level of monitoring and/or engagement in 2024. This is based on Ofwat's (the Water Services Regulation Authority) assessment of their financial position, challenges, and in some cases, the actions they need to take to address and strengthen their financial resilience, including to ensure they are appropriately financed to deliver their AMP 8 (Asset Management Plan 2026–31) investment plans (Ofwat, 2024a, p. 4).

According to a water company performance report 2023–24, companies need to take action to reduce pollution incidents that have increased in 2023 and now are at an unacceptable level. Almost all companies reported an increase in internal sewer flooding incidents in 2023–24. Most companies reported a reduction in annual leakage but improvements still need to be accelerated. Customer satisfaction has continued to fall and is now at its lowest since this performance measure was introduced in 2020–21. The majority of companies improved their performance for water supply interruptions in 2023–24 but some still have unacceptable levels (Ofwat, 2024b).

4.2 LOCALITY OF WATER SERVICES

As discussed in the introduction to this chapter, *locality, collaboration and local democracy* are important in water services. It is worth keeping in mind the multi-level governance of water services and the diversity of types from property-managed systems, community- and consumer-managed cooperatives, municipal utilities to various types of intermunicipal systems and related institutional arrangements. In the USA, water and sanitary districts are mentioned as a special intermunicipal arrangement used also for irrigation (Melosi, 2008, p. 136). Other countries may have their own respective bodies and frameworks.

Peltomaa (2013) pointed out that water services become particularly attached to the locality. Therefore, it is often feasible to arrange water services on a local basis. Although automatisation and advanced technologies will provide additional opportunities, distances related to production need to be reasonable. As one water utility manager noted 'It is no use to sit in the car driving around all day long'.

Thackston *et al.* (1983, cited by Grigg, 1996, p. 442) evaluated water policy in Tennessee and found a list of constraints. They concluded that institutional issues are the greatest impediment to regionalisation and that the success of regionalisation is based on 70% politics, 20% engineering, and 10% on good luck. If this is valid more widely today, should for example, engineering education take this challenge more seriously?

Grigg of Colorado (Grigg, 1996, p. 442) and Lieberherr of Switzerland (Lieberherr, 2011) among others, pointed out that large units may not improve the safety of water supply services, but may increase vulnerability more than several smaller units. When a problem occurs in a small unit, its range of influence is narrower. If a failure occurs in a large system, it can have much wider impacts.

As early as 1973, in his classic work Small is Beautiful, economist E.F. Schumacher (1911–1977) considered how many different structures are needed for different activities (Schumacher, 1973, p. 70). Organisations can be small and large and there is no single answer to their appropriate size. Yet, according to him, it seems difficult for people to understand why there may be seemingly opposing solutions to a problem. In fact in spite of the title of his book, the major argument by Schumacher (1973) seems to be in favour of the diversity of size dependent on conditions rather than smallness (first author's note).

Schumacher (1973, p. 70) continued 'They always tend to clamour for a final solution, as if in actual life there could ever be a final solution other than death'. According to him there is an appropriate scale for every activity, so the more active and intimate the activity, the smaller the group. This is the case, for example, in small water cooperatives where an active promoter and often also the manager (the 'champion') plays an important role (Katko, 1994) – at least in cultures that emphasise individualism.

Chorn (2011) suggested that economy of scale has only a limited impact on people working in a knowledge-based service organisation. In centralising services, to save duplication, the complexity of the processes is increased. This causes an increase in failures, and results in higher total organisational costs.

In Finland, some small municipal utilities are willing to merge their water services with those of larger utilities or with regional companies. Their willingness to do this depends largely on existing conditions and their needs. Several intended mergers have never been realised, the sticking points being the value of the assets. The managers of a small municipality have to be experts in all matters, including those beyond their field of expertise. It can be a relief for small municipalities to hand over responsibility for water supply and sewerage management to people who are recognised experts in their specific field. Yet, some smaller utility experts stress the importance of all-round skills; all staff members know what to do when rapid repair is needed. Thus, it is obviously not a question only of size (Juuti & Rajala, 2013).

On the other hand, mergers of completely *separate businesses* such as energy and water works have not created synergy benefits as were argued by their promoters. These sectors have their special features. A plumber is not able to act as an electrician based on professional requirements and the vice versa. Water and electricity are seldom installed simultaneously, and individual properties need electricity and water connections at different times (Figure 4.12). In fact, water services seem to have more synergies with street constructions (Katko *et al.*, 2022b, pp. 140–142).

If water and energy utilities are allocated adequate autonomy including over their finances, merged utilities or conglomerates may be justified. If an electricity company as a bigger unit dominates operations, the professional competence of water experts is bypassed. This has happened sometimes in Finland due to a lack of proper control, although in several cases sensible cooperation and some synergies have been created. Such synergies are found more between water services and district heating. The issue of adequate autonomy is probably the key here (Katko *et al.*, 2022a, pp. 140–146). In the USA, cities often have Public Works Departments that may include autonomous water and wastewater, solid waste, and gas utilities (Grigg, 1996, pp. 195–200). Under the American federal system each level of government needs to fill essential roles in managing water uses for various purposes. According to Grigg (2022, p. 117) 'planning- and policy development are the major areas of needed state activity to help forge holistic solutions to regional water problems'.

In Helsinki, the capital of Finland, water supply and sewerage and waste management as well as regional and environmental information services were merged from the beginning of 2010 and are now known as the Helsinki Region Environmental Services Authority (HSY). It was driven by politicians with the views of the independent water utilities in four municipalities in the region completely ignored in decision making in spite of the fact that they all managed their services well (Katko, 2016, p. 135; Juuti & Rajala, 2013).

The cases of merging separate sectors described above are indicative of the ideas of New Public Management (NPM) that had a delayed start in Finland (Vinnari, 2008). Municipalities were allowed to use the financial management models and business bookkeeping of the private sector and to make long-term economic plans and budgets. Based on all public material and documents, it is obvious that the actual reason for the merger was metropolis policy and politics – not the need to develop water services (Jutti & Rajala, 2011, p. 185).

Instead of merging separate sectors, examples of multi-level governance of water services in Finland are discussed below. The first contract for bilateral water sales was signed in 1959 between Espoo and Kauniainen, next to Helsinki. In 1961, Helsinki began to receive wastewaters from Vantaa. This type of contractual cooperation tripled from 1975 to 2006, especially for wastewater treatment. Contractual cooperation often also promotes closer intermunicipal or regional cooperation (Kurki *et al.*, 2010, 2016).

The first Finnish municipal water federation was established in 1954 and the first regional water company in 1965. The first wholesale water company was born in 1968 and the first wholesale wastewater treatment company in 1971.

The first large water company in Hämeenlinna region was established in 2001 and is discussed below (Katko, 2016, p. 135).

This regional water and wastewater company was established for the region exceptionally quickly. Prior to this, municipalities had positive experiences of good cooperation in the IT and education sectors. An effective tactic was to give the neighbouring municipalities a majority on the board to make sure they would not feel overruled by the central city (Juuti & Katko, 2005, pp. 64–67). Municipal mergers that happened later reduced the number of board members. According to Heinonen (2010), decisions were made at the right time to make use of the political momentum. This is something that does not necessarily exist elsewhere. Another regional water and wastewater company was established in southeastern Finland in 2007 (Pietilä *et al.*, 2010, p. 95). In general, inter-municipal, supra-municipal and regional water entities may be connected to each other in different ways (Katko, 2016, p. 135).

REALITY: In some cases, the size of a WWS utility can be increased if there is sufficient local knowledge. Above all, cooperation between organisations should be promoted. Combining small utilities does not automatically eliminate water supply problems, such as underinvestment in network renovation and renewal, and a lack of proactive communication between stakeholders. Merging water with energy is recommended only if adequate autonomy of both sectors can be assured.

In Kenya, the water services regulator recommends merging commercially viable and sustainable utilities rather than using mergers to create commercial viability and sustainability (WASREB, 2021). This recommendation was made after various merged county utilities opted to unmerge and community water projects seemed to be opposed to merging with these utilities. For example, in the court case referenced George Ngotho & 26 others v Governor of Kiambu County & 6 others (Kenya Law Review, 2019), community water projects challenged the county government's decision to take control of them., as they prefer autonomy. Nonetheless, as the government sees merging to be beneficial for regulation and economies of scale, a merger strategy would, at best, involve a careful negotiation to achieve a win-win situation rather than a forceful takeover.

In merging smaller units there will be an ultimate requirement to control the new larger unit through more rigid and formal procedures, including the increasing need for reporting. Thus, a small and flexible unit that is felt as one's own, is replaced by a more rigid system. Although a bigger unit may in many cases be advantageous, it also has its negative aspects. As discussed in Myth 4, the integration of water and wastewater utilities has many merits.

Regarding organisational options for producing water services, it is good to bear in mind the words of the late Professor Matti Viitasaari (1932–2000): 'Man has not invented a type of organisation that would hinder cooperation as such, if sensible people are willing to cooperate'.

Merging of the utilities alone is not sufficient measure to ensure anticipated and desirable positive outcomes, since it is clearly shown above that even large utilities can face difficulties. North (1993, p. 7) underlined that in addition to formal and informal institutions – the rules of the game – the *enforcement characteristics* of both have also to be considered and included. The predominant rules in society are functioning correctly only when they are enforced comprehensively and their regulatory compliance is monitored, promoted and guaranteed. Therefore, effective mechanisms for ensuring compliance with rules and regulations, and in addition, their correction procedures, are necessary.

Myth 17: Water services should be privatised to make them more efficient and to promote competition

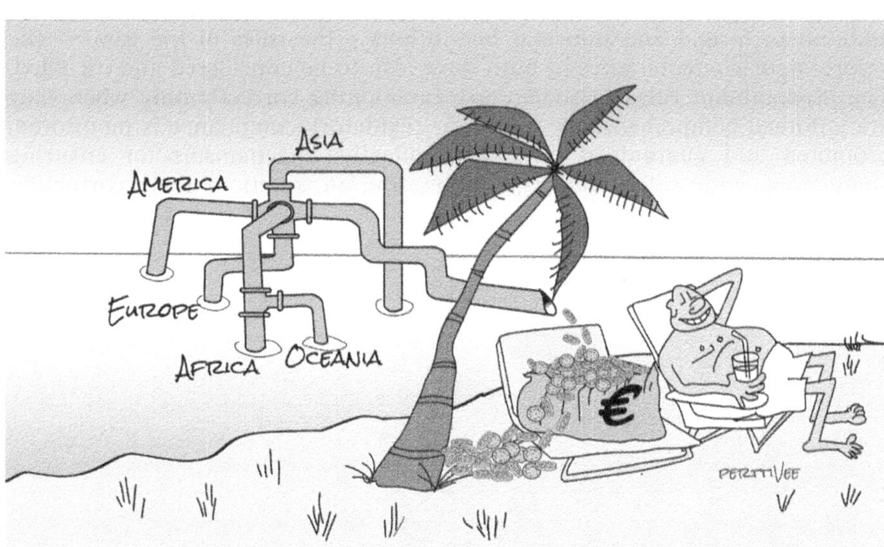

Figure 4.13 Is this how competition works in a market economy?
Illustration: Pertti Väyrynen.

In many European countries and especially in North America, the first modern water systems were built as private systems, which in most cases were taken over by municipalities by 1900 (Figure 4.13). Water services privatisation started to gain ground again in the 1980s due to the idea of New Public Management (NPM) until the full privatisation of water services was carried out in England and Wales in 1989. A more recent development is the return to municipal control of water services (Katko, 2016, pp. 206–208).

Worldwide, private enterprises have two key fundamental roles in water utility services production (Figure 4.14, from left to right):

(1) *Delivery of specific services, goods and civil works contracts*: mechanical, electrical, information technology and artificial intelligence installations and systems, that is design and planning, civil works construction, maintenance, supplying equipment and spare parts, and various specific consultancy, research, development and innovation services; option (a), on the left, is most commonly used worldwide.

(2) *Management of the whole or part of the utility's water services production*, that is operating the utility or its key facilities or components – for example, networks, pumping stations or treatment plants – based on the various types of service production contracts of different durations. The most essential options are: (b) Management and operating contracts (3–5 years); (c) Lease and affermage contracts (8–15

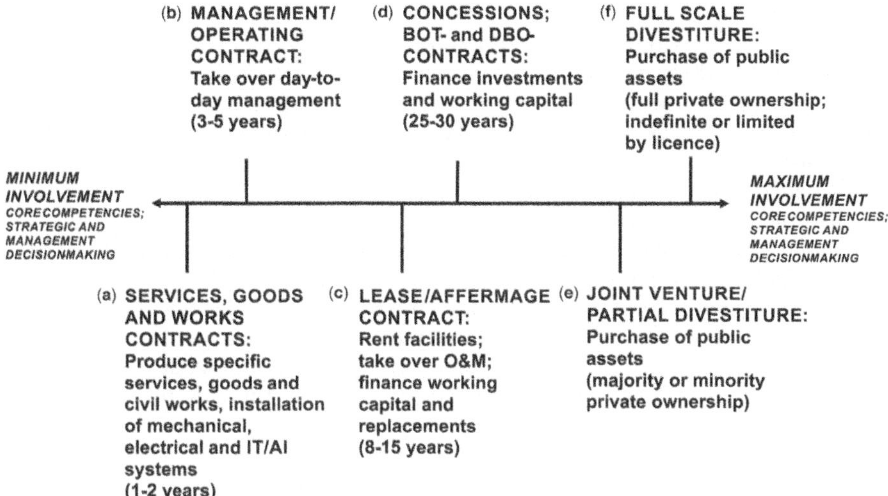

Figure 4.14 Spectrum of private sector arrangements in water services.
Source: WASH (1992); Delmon (2010, p. 14), modified by Hukka and Katko (2025).

years); (d) Concession, Build-Operate-Transfer (BOT) and Design-Build-Operate (DBO) contracts and other project delivery models modified from the BOT model (25–30 years); (e) Forming a joint venture with the public owner or having majority or minority ownership of the company; and (f) Owning the company fully (indefinitely or with the duration limited by licence).

In this context, options (b) to (f) are categorised under the definition 'privatisation'. It is, however, of utmost importance to remember that the majority of the world's urban population relies on public sector water services. For instance, in OECD countries, approximately 75–83% of residents were estimated to be connected to public utilities in 2009. Private water supply is predominant (i.e., serves over 50% of the population) in only five countries of the world, three of which belongs to the OECD: Chile, the Czech Republic, France, Malaysia and England. Conversely, private sector participation is low or non-existent (serves less than 10% of the population) in 17 out of the 30 OECD countries (Pérard, 2009, p. 194).

In 2007, public sector supply in developing countries and emerging economies was even higher, accounting for 93% of the total urban population (Marin, 2009). In mid-2006, in middle- and low-income countries, 90% of the largest cities – with a population of over one million – were served by a public sector operator. At the time, this public sector dominance was growing, as private companies retreated from many of the concessions and leases in developing countries. In rural areas, where there is little profitable business for private companies, the percentage of water services provided by the public sector was even higher – at least 95% (Hall & Lobina, 2006).

According to PSIRU (2014, p. 4), it is often assumed falsely that privatisation or public-private partnerships (PPPs) result in greater levels of technical efficiency. Therefore, one of the main arguments of privatisation proponents has been that it can always deliver a given level of service with less input costs than the public sector. Politicians, media, academics and consultants frequently refer to 'private sector efficiency'. Pérard (2009, p. 193) argued that most theories on private sector participation in water infrastructure are based on the sole assumed difference in efficiency between the public and private sectors. However, a review of 22 empirical tests and 51 case studies shows that private sector participation per se in water supply does not systematically have a significant positive effect on efficiency. The evidence for developing countries shows the same picture. A World Bank paper in 2005, reviewing studies on the water industry worldwide, concluded that 'the econometric evidence on the relevance of ownership suggests that in general, there is no statistically significant difference between the efficiency performance of public and private operators in this sector' (Estache & Goicoechea, 2005; PSIRU, 2014, p. 21).

In Paris, the *remunicipalisation* of water supply ceased 25 years of private water management in January 2010. The reasons that convinced the city of Paris to remunicipalise water supply included price hikes that were not justified by private operator investments, the lack of financial transparency under private management, difficulties in achieving sustainable development objectives through the renegotiation of private contracts, and the greater degree of public control and managerial flexibility that come with public ownership. In addition, it was found that the price charged by the two private operators in Paris was 25% to 30% higher than was economically justified. Remunicipalisation has enhanced sustainable water development financially, economically, technically, politically, socially and environmentally. It has produced transformative results at various levels due to a policy that has shifted the interests to consumers, the environment and the local community. Remunicipalisation has enabled Eau de Paris to enjoy low indebtedness, high self-financing ratios and high investment levels. Furthermore, Eau de Paris has increased efficiency and reduced prices, inverting the 25-year trend of above-inflation price increases under private management (Lobina *et al.*, 2021, pp. 7–9).

In England and Wales, the Thatcher government privatised ten publicly owned Regional Water Authorities (nine in England and one in Wales) in November 1989. The privatisation process transferred ownership of the water and sewerage operators to ten new companies, listed on the stock exchange (Hall & Lobina, 2024, pp. 5–6). A series of studies in the UK has found that there has not been any significant improvement in productivity performance since privatisation; a 2007 report concluded that: 'after privatization, productivity growth did not improve ... average efficiency levels were actually moderately lower in 2000 than they had been at privatization in 1989' (Hall, 2014).

Yearwood (2018, p. 2) argued that in England and Wales the 40% increase in real household bills since privatisation was mainly driven by continuously growing interest payments on debt, contrary to the regulator (Ofwat) attributing them to growing costs and investments. The accelerating debt levels were primarily the result of disproportionate dividend pay-outs, which has exceeded privatised companies' cash balances in all but one year since 1989. He concluded that the way the industry operated may no longer be sustainable and seems to greatly disadvantage consumers without their knowledge, due to the fog of misleading statements by the companies and Ofwat. Furthermore, customers in England seem to be losing from this privatised industry when benchmarked with publicly owned Scottish Water. The efficiency and quality of the English industry is on a par with Scotland, although investments are lower and costs to consumers are much higher (Yearwood, 2018, pp. 25–26).

According to Lobina (2013), the limited competitiveness of the water services sector is due to the market structure and private operators' interest seeking behaviour. Opportunism also allows private operators to appropriate net gains when interacting with contract awarding and regulatory authorities under different institutional frameworks. If private operators obtain productive efficiency, power differentials allow them to retain this as rent rather than passing it onto consumers.

The empirical evidence undermines a fundamental part of the argument in favour of privatisation and use of the private sector. If private companies are not more efficient, then the most common rationale for privatisation is not valid. The reason for this is that privatisation, outsourcing and public-private partnerships (PPPs) are facing a clear disadvantage in relation to most other economic criteria. The major disadvantage is that the cost of investment finance is nearly always significantly more expensive for private operators, due to higher profits for shareholders and lower credit ratings, which means that private companies pay higher interest rates. Unless the private sector can deliver substantial savings from efficiency, it is more expensive than the public sector. Secondly, efficiency is not the same as cutting costs, since lower costs may simply mean lower quality of service. A private company may also take its profits by cutting the jobs, pay and conditions of its workers, without improving the systems of work and efficiency. The public sector carries the extra 'transaction costs' of sales, tendering, monitoring and regulation. A low tender may be used to win a contract, although the private contractor then renegotiates the price upwards – or the quality downwards – to become more profitable (PSIRU, 2014, pp. 4–5).

An assessment of global public spending offers an accurate estimate of public spending in the water sector in 130 countries covering four subsectors – water supply and sanitation (WSS), irrigation, water transport, and hydropower. Patterns in project-level infrastructure investment, once again, affirm the public sector as the prime source of WSS infrastructure development investment, and the limited role played in this by the private sector. In 2017, 91% of investments were from the public sector and just 9% from the private sector. Between 2009 and 2019, private sector investment projects made up only about 8% of the total number of WSS infrastructure projects (Joseph *et al.*, 2024, p. xxxvi). According to Hayward (2024), these rates are not expected to change in the future.

In England, Yearwood (2018 cited by Hall & Lobina, 2024, p. 14) found that in almost every year since privatisation, consumers' bills directly covered the capital expenditure (capex) of the water companies, as well as the day-to-day operational expenditure, with a small surplus. Despite this, companies have continued to borrow, building up a total debt of GBP 64.4 billion by 2023. This was not, however, needed to finance capex, but was used to pay out much larger dividends. Therefore, shareholders have not been paying capex in England since 1990, rather, customers have been financing this (Hall & Lobina, 2024, p. 14).

Before privatisation, all the water companies were debt free. These debts have not deterred companies from paying dividends to shareholders. In 2022–23, a total of GBP 1.446 billion was paid out in this way. This is equivalent to over GBP 53 per customer in each household on the revenue made through bill payment (Hall & Lobina, 2024, p. 28). The largest dividend payments were made by Severn Trent and United Utilities, representing 22% and 25% of their entire revenues respectively. All the companies themselves pay very little tax on their profits: a total of just GBP 45 million in 2022–23 which is equivalent to just 3% of payments to shareholders (Hall & Lobina, 2024, p. 12). Further, Anglian, Southern and Thames paid no corporation tax in 2018, while Thames has paid no corporation tax for a decade (Plimmer & Pickard, 2018). Companies have also borrowed money to pay dividends, with the cost of servicing these debts being additional to the cost of the dividends themselves. When the estimated annual costs of the debt interest are added to the cost of the dividends, the total represents about 29% of total annual revenues (Hall & Lobina, 2024, p. 12).

Out of the 16 largest water companies in England and Wales, ten are categorised as requiring an increased level of monitoring and/or engagement in 2024 (Ofwat, 2024a).

REALITY: The private sector has not been proven to be more efficient than the public sector. It does not provide its own funds to finance the investments – instead it uses the same sources as the public sector. Private investments, ownership and management become more expensive than public investments, ownership and management. Neither does privatisation increase competition and, in fact, privatised water services can reduce competition when private companies win long-term contracts and supply equipment from their own subsidiaries. Privatisation also opens up the markets for competition for natural monopolies. Globally, there has been an emerging trend to remunicipalise water services. This is in many cases a response to the empty promises of private companies and their failure and unwillingness to respect the needs of the communities and environmental concerns.

In the USA, privatisation and so-called 'managed competition' also received wide interest in the late 1990s (APWA, 1998). However, private contracts were soon cancelled in many of the cities concerned in the early 2000s.

In Estonia, the city of Tallinn sold 50.4% of the shares in its water services company to international investors in order to secure investment funds in 2001. In the following years, the city's expectations of privatisation were not met. The private owner extracted large sums of money from the company, while investments were financed with loans from the European Bank of Reconstruction and Development (EBRD). The actual investment levels and those agreed in the business plan were much lower than the need assessments used to support the privatisation, and AS Tallinna Vesi had the responsibility of implementing these investments, rather than the private owner. The stipulations of the Service Agreement proved to be unclear and contrary to the spirit of the original tender documents that aimed to deliver stable tariffs. The city ended up paying for stormwater services as well as the connection fees of new customers. Water tariffs were also considerably increased (Vinnari & Hukka, 2007).

In 2005, AS Tallinna Vesi was listed in Nasdaq Tallinn, and the private owner United Utilities received EUR 27.2 million from selling shares. In February 2021, United Utilities agreed to sell its 35.3% stake in AS Tallinna Vesi for EUR 100.26 million to a consortium of the city of Tallinn and the energy group OÜ Utilitas (Nasdaq, 2021). Thus, the net profit of the private owners was EUR 86.46 million from selling these shares of the privatised company which had been solely owned by the city of Tallinn prior to 2001.

Contrary to expectations, privatisation has not relieved governments of the burden of investment financing. Private water companies do not bring new sources and volumes of investment finance – they rely heavily on the same sources as are available to the public sector. This evidence undermines one of the most important myths concerning water privatisation, namely that private finance plays an important role in delivering progress for water and sanitation coverage. On the contrary, this has not occurred up to now and is unlikely to do so in the future (Hall & Lobina, 2006, p. 51; Araral, 2009). Furthermore, according to Bayliss and Hall (2002), private financial institutions cannot take the risks that developing economies present without some kind of guarantees and insurance, which are borne by taxpayers and consumers in the South.

In addition, privatisation is extremely complicated since there are high transaction costs (Lamberg *et al.*, 1997). These accrue from the time and efforts used for planning, decision making, negotiations, preparing and enforcing contracts, arbitration and dealing with breaches of contract, coordination and other related activities (Oulasvirta, 1994). Transaction costs are not insignificant and may vary considerably depending on the institutional setting. For example, Ofwat – the economic regulator of the water sector in England and Wales – is financed by licence fees recovered from the privatised water companies. Ofwat's budget for 2017–18 exceeded GBP 25 million and spending in 2014–15 was over GBP 29 million (Lobina, 2018).

According to Prasad (2006), it has been argued that private sector participation (PSP) in water supply would help the poor to access services at an affordable price. Large multinational companies are not interested in low-income countries where there is lack of commercial viability of water supply. From the private sector's perspective, low-income countries and the poor in particular, are unattractive and present high levels of risk. In order to lower this risk, the private sector 'cherry picks' the better-off customers in urban areas or less risky environments. It may alternatively rely on subsidies, soft loans, and a renegotiation of the contractual agreement in order to provide services to the poor. Private companies use the same sources of funds as the public sector, namely, loans from bilateral and multilateral donors, aid money and revenue from customers through tariffs. Therefore, public funds are used to support the private sector in providing services to the poor (Global Water Intelligence, 2005 cited by Prasad, 2006). The World Bank has also clearly stated that even where there is PPP, public funding is essential to meet the required investments (Harris & Janssens, 2004, pp. 8, 15).

In France, private water service operators were established in the mid-1800s and have gradually captured markets around the world in this and other service sectors. In practice, these operators have received concessions of up to 30 years to manage facilities, resulting in hardly any competition, particularly when goods, services and works needed by a utility are procured from the operator's subsidiaries. Competition for a new concession or lease (affermage) is false since only the enterprise itself knows the facility well enough to be able to minimise risks. According to Mattisson (1993), new French contracts have been granted almost without exception to the previous operators. Moreover, three companies have in practice divided the market geographically (Hukka & Katko, 2003). Multinationals have also established joint ventures between themselves (Hall, 2002) making the issue of competition even more questionable. As they weakened financially in the late 2000s, all the companies became vulnerable to takeovers and a national strategy to protect key French companies from foreign control resulted in the state becoming the major shareholder in all three. This resulted in the partial nationalisation of these private French companies (Hall & Lobina, 2012a, p. 128).

Hall and Lobina (2004) concluded that competition cannot be successfully introduced, and that the effectiveness of instruments such as regulation is frequently limited by the bargaining asymmetry related to superior legal, technical and economic resources enjoyed by private water companies vis-à-vis public authorities, especially when these are in low-income countries. Hall and Lobina (2012b) argued that private business in this sector has been built by gaining uncompetitive and sometimes corrupt monopolies, and then holding onto them for decades, in some cases for over a century (Table 4.2).

In Spain, two-thirds of private contracts have never been subject to competition, and three cities – Barcelona, Valencia, Alicante – have private water operations which have run for over 100 years without any competition. The case of Barcelona is remarkable because there is no legal concession contract according to a recent court ruling (Hall & Lobina, 2012a, p. 11).

Table 4.2 Private water contracts – no competition.

	Proportion of Private Water Contracts Based on Competition
France	5%
Spain	33%
UK	0%

Worldwide, there has been a noticeable trend of cities, regions and countries increasingly choosing to remunicipalise services by taking back public control over water and sanitation management. Since 2000 to September 2024, there have been no less than 377 cases of water remunicipalisation globally (Public Futures, 2024). This term refers to the return of previously privatised water supply and sanitation services to public service delivery and for water, this is more than a mere change in ownership of service production; it also represents a new possibility for the realisation of collective ideas of development, such as the human right to water and sustainable water development. In other words, remunicipalisation offers opportunities for building socially desirable, environmentally sustainable, quality public water services benefiting present and future generations (Kishimoto *et al.*, 2015). Based on long-term experience of and findings on privatisation, Lobina *et al.* (2021, p. 23) pointed out that it is better not to privatise than to remunicipalise and then face the costs of contract termination.

Hontelez (2002) argued that the way forward in sustainable water management in the European Union member countries is to improve existing structures of water supply and sanitation services – not to sell and liberalise water services. Major reforms are necessary and should focus on four areas: *public participation, resource protection, highest standards and sound water prices.* The lobbying and pressure to liberalise and commercialise European water services is not driven by these kinds of intentions, but rather by the desire to increase market access by the multinationals and to find short-term financial resources for weak state budgets. Therefore, Hontelez (2002) suggested the following, with which the authors concur:

- Full democratic control of water resources, water supply and sanitation and their long-term stability and safety is most likely to be achieved under public ownership.
- Constant improvement of current public systems is necessary and possible under public auspices.
- All potential organisational forms of water services, especially public-public partnerships, must be considered.
- Operation of systems by public-private partnerships is welcome if they respect full public control over strategic decisions.

4.3 DISCUSSION QUESTIONS

(1) What principles relating to successful soccer can be applied to water services and collaboration?
(2) Why do some people assume that water and sanitation as a basic human right equates to free services?
(3) What is the value of water in society?
(4) How should technology be understood and interpreted?
(5) Where do the arguments about automatic benefits of economies of scale originate?
(6) Why do we need tax havens?

REFERENCES

Andrews M. (2013). The Limits of Institutional Reform in Development: Changing Rules for Realistic Solutions. Cambridge University Press, Cambridge, UK, p. 82.

Antila K., Katko T. S. and Mattila H. (2013). Technology development theories and water services evolution. In: Water Services Management and Governance: Past Lessons for a Sustainable Future, T. Katko, P. Juuti and K. Schwartz (eds), IWA Publishing, London, pp. 13–26, https://library.oapen.org/handle/20.500.12657/25810 (accessed 22 March 2025).

APWA (American Public Works Association). (1998). Congress, 13–19 Sept 1998, Las Vegas, Nevada.

Araral E. (2009). The failure of water utilities privatization: Synthesis of evidence, analysis and implications. *Policy and Society*, **27**(3), 221–228, p. 226, https://doi.org/10.1016/j.polsoc.2008.10.006

Bauby P., Hecht C. and Warm S. (2018). Water remunicipalisation in Berlin and Paris: specific processes and common challenges. CIRIEC No. 2018/07, https://www.ciriec.uliege.be/wp-content/uploads/2018/10/WP2018-07.pdf (accessed 22 March 2025).

Bayliss K. and Hall D. (2002). Can Risk Really be Transferred to the Private Sector? A Review of Experiences with Utility Privatisation. Public Services International Research University. University of Greenwich, Greenwich, UK, p. 9.

Becker J. (1993). Survey says water utility advisory councils are a success. *Journal AWWA*, **85**(11), 58–61, https://doi.org/10.1002/j.1551-8833.1993.tb06101.x

Behailu B. M., Pietila P. E. and Katko T. S. (2016). Indigenous practices of water management for sustainable services: case of Borana and Konso, Ethiopia. SAGE open, https://doi.org/10.1177/2158244016682292

Beilinskij A. (2015). Vesihuoltolakiopas 2015 (The guidelines for the Water Services Act). Ministry of Agriculture and Forestry Publication 5/2015, https://mmm.fi/documents/1410837/1720364/MMM_5_2015.pdf/383bfb97-d522-49de-9602-46fbb958cb4a (accessed 22 March 2025).

Benöhr I. (2023). The Right to Water and Sustainable Consumption in EU Law. *Journal of Consumer Policy*, **46**, 53–77, https://doi.org/10.1007/s10603-022-09532-5

Berg S. (2013). Best Practices in Regulating State-owned and Municipal Water Utilities. Economic Commission for Latin America and the Caribbean, CEPAL, p. 7, https://repositorio.cepal.org/entities/publication/799eaece-ee17-434e-97ce-ca71170b4ca1 (accessed 22 March 2025).

Berninger K., Laakso T., Paatela H., Virta S., Rautiainen J., Virtanen R., Tynkkynen O., Piila N., Dubovik M. and Vahala R. (2018). Tulevaisuuden kestävä vesihuolto – ennakointi, ohjaus ja järjestäminen (Sustainable water services for the future – direction, steering and organisation). Valtioneuvoston selvitys ja tutkimustoiminnan julkaisusarja 56/2018. Valtioneuvoston kanslia. Helsinki, http://urn.fi/URN:ISBN:978-952-287-607-2 (accessed 22 March 2025).

Bijker W., Hughes T. P. and Pinch T. (eds) (1987). The Social Construction of Technological Systems: New Directions in the Sociology and History of Technology. MIT Press, Cambridge, MA, USA.

Bluechain Consulting. (2024). Lessons from the Privatisation of UK Water Companies – Challenges and Responsibilities, https://www.bluechainconsulting.com/post/lessons-from-the-privatisation-of-uk-water-companies-challenges-and-responsibilities (accessed 22 March 2025).

Böckenförde M. (2011). Decentralization from a Legal Perspective: Options and Challenges. Gießen, TransMIT, p. 4, https://www.idea.int/sites/default/files/publications/chapters/practical-guide-to-constitution-building/a-practical-guide-to-constitution-building-chapter-4.pdf (accessed 22 March 2025).

Cardone R. and Fonseca C. (2003). Financing and Cost Recovery. Thematic Overview Paper No. 7. IRC International Water and Sanitation Centre, p. 25, https://sswm.info/sites/default/files/reference_attachments/CARDONE%20and%20FONSECA%202004%20Financing%20and%20Cost%20Recovery.pdf (accessed 22 March 2025).

Chorn N. (2011). Economy of Scale is a myth. Why shared services don't deliver. Brainlink group, https://www.drnormanchorn.com/economy-of-scale-is-a-myth (accessed 22 March 2025).

Cox M., Gwen A. and Villamayor-Tomás S. (2009). A Review and Reassessment of Design Principles for Community-Based Natural Resource Management. Unpublished.

Cromartie J. and Vilorio D. (2019). Rural Population Trends. Statistic: Rural Economy & Population, February 15, 2019, https://www.ers.usda.gov/amber-waves/2019/february/rural-population-trends/ (accessed 22 March 2025).

Davis K.E. (2009). Institutions and Economic Performance: An Introduction to the Literature. New York University Law and Economics Research Papers, No. 09–51, New York. https://ssrn.com/abstract=1520515 (accessed 22 March 2025).

Delmon J. (2010). Understanding options for public-private partnerships for infrastructure: sorting out the forest from the trees: BOT, DBFO, DCMF, concession, lease … Policy Research Working Paper 5173, World Bank, Washington, DC.

Estache A. and Goicoechea A. (2005). A 'research' database on infrastructure economic performance (English). Policy Research Working Paper WPS3643, https://documents.worldbank.org/en/publication/documents-reports/documentdetail/823331468328564333/a-research-database-on-infrastructure-economic-performance (accessed 22 March 2025).

European Federation of National Associations of Water and Wastewater Companies (EurEau). (2016). Cost recovery and water pricing. Briefing. p. 1, https://www.eureau.org/resources/briefing-notes/144-cost-recovery-and-water-pricing-september2016/file (accessed 22 March 2025).

European Federation of National Associations of Water and Wastewater Companies (EurEau). (2017). Customers and cost recovery: realising the Water Framework Directive, p. 1, https://www.eureau.org/resources/position-papers/139-customers-and-cost-recovery-may2017/file (accessed 22 March 2025).

European Union. (n.d.). Facts and figures on the European Union, https://european-union.europa.eu/principles-countries-history/facts-and-figures-european-union_en (accessed 22 March 2025).

Farnault A. and Leflaive X. (2024). Cost recovery for water services under the Water Framework Directive. OECD Environment Working Papers No. 240. OECD Publishing, Paris, France, https://www.oecd.org/en/publications/cost-recovery-for-water-services-under-the-water-framework-directive_e2a363e3-en.html (accessed 22 March 2025).

Finland Population. (n.d.). Live, https://www.worldometers.info/world-population/finland-population/ (accessed 22 March 2025).

Gavin D. (2023). The Rural Migration Trend: What to Make of It, Why It's Happening and Where It's Headed. Associaton of Equipment Manufacturers (AEM), 16 Feb 2023, https://www.aem.org/news/the-rural-migration-trend-what-to-make-of-it-why-its-happening-and-where-its-headed (accessed 22 March 2025).

Global Commission on the Economics of Water. (2023). Turning the tide: a call to collective action, https://watercommission.org/wp-content/uploads/2023/03/Turning-the-Tide-Report-Web.pdf (accessed 22 March 2025).

Global Water Intelligence. (2005). Patience wins the day at Veolia. *Global Water Intelligence*, **6**(5), 10–12.

Global Water Partnership (GWP). (2000). Integrated Water Resources Management. GWP Technical Advisory Committee TAC Background Paper No. 4. Stockholm, Sweden, pp. 20–21, https://www.gwp.org/globalassets/global/toolbox/publications/background-papers/04-integrated-water-resources-management-2000-english.pdf (accessed 22 March 2025).

Grigg N. S. (1996). Regionalization in water management. In: N. S. Grigg, Water Resources Management. Principles, Regulations and Cases, McGrawHill, New York, pp. 441–460.

Grigg N. S. (2022). State government roles in water governance: time for an upgrade. Commentary. *Public Works Management & Policy*, **28**(2), 117–134, https://doi.org/10.1177/1087724X221146117

Haeffner M., Jackson-Smith D. and Flint C. G. (2018). Social position influencing the water perception gap between local leaders and constituents in a socio-hydrological system. *Water Resources Research*, **54**, 663–679, https://doi.org/10.1002/2017WR021456

Hall D. (2002). The Water Multinationals 2002 – Financial & Other Problems. PSIRU, University of Greenwich, London, p. 7.

Hall D. (2014). Public and private sector efficiency. A briefing to EPSU by PSIRU, p. 19, https://www.epsu.org/sites/default/files/article/files/Public%20and%20Private%20Sector%20efficiency%20EN%20fin.pdf (accessed 22 March 2025).

Hall D. and Lobina E. (2004). Private and public interests in water and energy. *Natural Resources Forum*, **28**, 268–277, p. 275, https://doi.org/10.1111/j.1477-8947.2004.00100.x

Hall D. and Lobina E. (2006). Pipe dreams: the failure of the private sector to invest in water services in developing countries. Public Services International Research Unit, University of Greenwich, London, UK, p. 3.

Hall D. and Lobina E. (2012a). The birth, growth and decline of multinational water companies. In: Water Services Management and Governance: Lessons for a Sustainable Future. Governance and Management for Sustainable Water Systems Series, T. Katko, P. Juuti and K. Schwartz (eds), IWA Publishing, London, UK, pp. 123–132.

Hall D. and Lobina E. (2012b). Utterly Uncompetitive – the Home Lands of Eternal Water Privatisations. PSIRU, University of Greenwich, London, p. 1, https://www.world-psi.org/sites/default/files/documents/research/psiru_water_competition_eu_short.pdf (accessed 15 Dec 2024).

Hall D. and Lobina E. (2024). Clean Water: A Case for Public Ownership. Presented on 16 June 2024 at UNISON Water, Environment and Transport (WET) Conference Fringe in Brighton, UK.

Harris C. G. and Janssens J. G. (2004). Operational guidance for World Bank Group staff: public and private sector roles in water supply and sanitation services. World Bank, Washington DC. http://documents.worldbank.org/curated/en/556271468155727649

Hatakka A. (2016). From global to local, from dependencies to growing self-sufficiency and from risk-taking to resilience: a qualitative content analysis on degrowth as an institutional change and a future contribution in Finland. MSc thesis, Environmental Policy and Regional Studies, School of Management, University of Tampere, Finland.

Hayward K. (2024). Service providers and the route to security. The Source, **35**, 10.

Heikkila T. (2004). Institutional boundaries and common-pool resource management: a comparative analysis of water management programs in California. *Journal of Policy Analysis and Management*, **23**(1), 97–117, https://doi.org/10.1002/pam.10181

Heino H. (2016). Paradigman jäljillä. Tutkimus vesihuollon ajattelumalleista (Exploring the Paradigm. Research on Thought Patterns of Water Services). Doctoral dissertation no. 1374, Tampere University of Technology, Tampere, Finland, https://trepo.tuni.fi//handle/10024/114521 (accessed 22 March 2025).

Heinonen T. (2010). 15 Jan 2010, Personal communication, Finland.

Heller L., Karunananthan M., Zwarteveen M., Hall D., Manahan M. A. and Diouf F. (2023). What water will the UN Conference carry forward: a fundamental human right or a commodity? *Lancet*, **402**, 757–759, https://doi.org/10.1016/S0140-6736(23)01430-7

Helsinki Term Bank for Arts and Science. (2024). Appropriate technology, https://tieteentermipankki.fi/wiki/Nimitys:appropriate_technology (accessed 22 March 2025).

Hendricks D. (2010). Fundamentals of Water Treatment Unit Processes: Physical, Chemical, and Biological. CRC Press, Taylor & Francis Group, Boca Raton, Florida, USA, p. 14.

Hendricks F., Loughin J. and Lidström A. (2011). Comparative reflections and conclusions, pp. 715–742. In: Local and Regional Democracy in Europe, J. Loughin, F. Hendricks and A. Lidström (eds), Oxford University Press, Oxford, UK, p. 727.

Hooghe L. and Marks G. (2002). Types of Multi-Level Governance. Les Cahiers européens de Sciences Po, n 03, p. 8, https://www.sciencespo.fr/centre-etudes-europeennes/sites/sciencespo.fr.centre-etudes-europeennes/files/n3_2002_final.pdf (accessed 22 March 2025).

Hontelez J. (ed.) (2002). Environmental Principles for Water Services in the EU: Private Versus Public Water Services. Position Paper of the European Environmental Bureau (EEB). EEB Publication Number: 2002/006, pp. 19–20.

Hughes T. P. (1987). The Evolution of Large Technological Systems. In: The Social Construction of Technological Systems: New Directions in the Sociology and History of Technology, W. Bijker, T. P. Hughes and T. Pinch (eds), MIT Press, Cambridge, Massachusetts, pp. 51–82.

Hukka J. J. and Katko T. S. (2003). Water privatisation revisited: Panacea or pancake? IRC Occasional Papers Series 33. Delft, The Netherlands, p. 127.

Hukka J. J. and Katko T. S. (2015). Appropriate pricing policy needed worldwide for improving water services infrastructure. *Journal AWWA*, **107**(1), E37–E46, https://doi.org/10.5942/jawwa.2015.107.0007

Hukka J. J. and Katko T. S. (2021). Towards Sustainable Water Services: Subsidiarity, Multi-level Governance and Resilience for Building Viable Water Utilities. CADWES Publications, Tampere University, Tampere, Finland. http://www.cadwes.com/publications/others/ (accessed 22 March 2025).

Hukka J. J. and Seppälä O. T. (2004). D10b: WaterTime National Context Report – Finland, www.watertime.net/wt_cs_cit_ncr.html# (accessed 22 March 2025).

Hukka J. and Katko T. (2007). Vesihuollon haavoittuvuus (Vulnerability of water services infrastructure). Foundation for Municipal Development. Publ. No. 58. https://kaks.fi/uutiset/vesihuollon-haavoittuvuus/ (accessed 22 March 2025).

Hukka J. J, Katko T. S., Mattila H. E., Pietilä P. E., Sandelin S. K. and Seppälä O. T. (2007). Inadequacy of positivistic research to explain complexity of water management. *International Journal of Water*, **3**(4), 425–444, https://www.inderscienceonline.com/doi/abs/10.1504/IJW.2007.016325, https://doi.org/10.1504/IJW.2007.016325

Hukka J. J., Castro J. E. and Pietilä P. E. (2010a). Water, policy and governance. *Environment and History*, **16**, 235–251, https://doi.org/10.3197/096734010X12699419057377

Hukka J., Katko T., Pietilä P., Seppälä O. and Vinnari E. (2010b). Forgotten infrastructure – the quest for development, sustainability and security. Security in Futures, Security in Change, 3–4 June, 2010, Turku, Finland, p. 322, https://www.researchgate.net/publication/352553735_Forgotten_infrastructure-the_quest_for_development_sustainability_and_security (accessed 22 March 2025).

Hynynen E.-L. (2011). Hannes Manninen 'Suuruuden ekonomia lyö kuntalaisia kasvoille'. *Polemiikki*, **15**(2), 8–11, https://kaks.fi/wp-content/uploads/2011/04/201102.pdf (accessed 22 March 2025).

Inha L. M., Katko T. S. and Rajala R. P. (2019a). Improved water services cooperation through clarification of rules and roles. *Water*, **11**(10), 2172, https://doi.org/10.3390/w11102172

Inha L., Juuti P., Katko T., Pietilä P. and Rajala R. (2019b). Vetoa vesihuoltopalveluihin täydennyskoulutuksen kautta (Driving continuing education in water services). *Kuntatekniikka*, **73**(5), 6–7, https://view.creator.taiqa.com/kuntatekniikka/kuntatekniikka201905#/page=6 (accessed 22 March 2025).

International Conference on Water and the Environment (ICWE). (1992). Dublin Statement on Water and Sustainable Development, https://ielrc.org/content/e9209.pdf (accessed 22 March 2025).

International Law Association (ILA). (2004). Water Resources Law. Berlin Conference 2004, https://internationalwaterlaw.org/documents/intldocs/ILA_Berlin_Rules-2004.pdf (accessed 15 Dec 2024).

Joosse H. and Teisman G. (2021). Employing complexity: complexification management for locked issues. *Public Management Review*, **23**(6), 843–864, https://doi.org/10.1080/14719037.2019.1708435

Joseph G., Hoo Y. R., Wang Q., Bahuguna A. and Andres L. (2024). Funding a Water-Secure Future: an Assessment of Global Public Spending. World Bank, Washington, DC, https://www.worldbank.org/en/topic/water/publication/funding-a-water-secure-future (accessed 22 March 2025).

Juuti P. and Rajala R. (2011). Veden vai metropolipolitiikan ehdoilla? (On the terms of water services or metropoly politics). HSY Veden syntyprosessi ja sen taustat. Tampub, Tampere, Finland, p. 185. https://trepo.tuni.fi/handle/10024/100502 (accessed 22 March 2025).

Juuti P., Harri Mattila H., Rajala R., Schwartz K. and Staddon C.(eds.) (2019). Resilient water services and systems: the foundation of well-being. IWA Publishing, London, UK. https://iwaponline.com/ebooks/book/764/Resilient-Water-Services-and-Systems-The

Juuti P. and Katko T. (eds) (2005). Water, Time and European Cities. History Matters for the Futures. Tampere University Press, ePublications, Tampere, Finland, pp. 64–67. https://trepo.tuni.fi/handle/10024/65706

Juuti P. and Rajala R. (2013). Reforming water services in the Helsinki metropolitan area. In: Water Services Management and Governance, P. Juuti, T. Katko, K. Schwartz, and R. Rajala (eds), IWA Publishing, London, pp. 179–188, https://library.oapen.org/handle/20.500.12657/25810 (accessed 22 March 2025).

Juuti P. S., Katko T. S. and Vuorinen H. S. (2007). Epilogue: Local Solutions Based on Local Conditions. In: Environmental History of Water – Global Views on Community Water Supply and Sanitation, P. S. Juuti, T. S. Katko and H. S. Vuorinen (eds), IWA Publishing, London, UK, pp. 593–598.

Katajamäki H. (2011). Paikallisyhteisöjen Suomi (Finland, a country of local communes). *Tiedepolitiikka*, **36**(3), 52–54.

Katko T. (1994). The need for 'champions' in rural water supply. *Waterlines*, **12**(3), 19–22, https://doi.org/10.3362/0262-8104.1994.007

Katko T. S. (2016). Finnish Water Services – Experiences in Global Perspective. Finnish Water Utilities Association, Helsinki, Finland. Co-published E-book, IWA Publishing, London, 2017, www.finnishwaterservices.fi (accessed 15 Dec 2024).

Katko T. S., Kurki V. O., Juuti P. S., Rajala R. P. and Seppälä O. T. (2010). Integration of water and wastewater utilities: a case from Finland. *Journal AWWA*, **102**(9), 62–70, https://doi.org/10.1002/j.1551-8833.2010.tb10187.x

Katko T. S., Juuti P. S., Pietilä P. E. and Rajala R. P. (2015). Water Services Heritage and Institutional Diversity. In: Water and Heritage. Material, Conceptual and Spiritual Connections, J. H. W Willems and H. van Schaik (eds), Sidestone Press, Leiden, the Netherlands, pp. 297–312, https://www.sidestone.com/books/water-heritage (accessed 22 March 2025).

Katko T. S., Juuti P. S, Juuti R. P. and Nieler E. J. (2022a). Managing Water- and Wastewater Services in Finland, 1860–2020 and beyond. *Earth*, **3**, 590–613, https://doi.org/10.3390/earth3020035

Katko T. S., Juuti P. S., Juuti R. P. and Väyrynen P. O. (2022b). Vesihuollon myytit (The Myths of Water Services). Vastapaino, Tampere, Finland.

Kemira. (2020). Highlights from an international consumer survey on water, https://www.kemira.com/app/uploads/2020/10/Kemira_water_datasummary_US_FINAL-5f9bf7b272098.pdf (accessed 22 March 2025).

Kenton W. (2024). Maastricht Treaty: Definition, Purpose, History, and Significance. 31 July 2024, https://www.investopedia.com/terms/m/maastricht-treaty.asp (accessed 22 March 2025).

Kenya Law Review. (2019). George Ngotho & 26 others v Governor of Kiambu County & 6 others [2019] KEHC 2180 (KLR).

Kishimoto S., Lobina E. and Petitjean O. (eds) (2015). Our Public Water Future: The Global Experience with Remunicipalisation. Transnational Institute (TNI), Public Services International Research Unit (PSIRU), Multinationals Observatory, Municipal Services Project (MSP) and the European Federation of Public Service Unions (EPSU). Amsterdam, London, Paris, Cape Town and Brussels, pp. 6–7, https://www.tni.org/en/publication/our-public-water-future (accessed 22 March 2025).

Kranzberg M. (1986) Technology and history: 'Kranzberg's laws'. *Technology and Culture*, **27**(3), 544–560.

Kuivamäki R., Kuulas A., Makkonen E., Renko T. and Laitala R. (2023). Selvitys vesihuollon organisoinnista (A study on organising water services). Monistesarja nro 87, Vesilaitosyhdistys, Helsinki, p. 39.

Kuntaliitto. (2024). Kaupunkien ja kuntien lukumäärät ja väestötiedot (Amounts of cities and their population), https://www.kuntaliitto.fi/kuntaliitto/tietotuotteet-ja-palvelut/kaupunkien-ja-kuntien-lukumaarat-ja-vaestotiedot (accessed 22 March 2025).

Kurki V. O., Katko T. S. and Pietilä P. (2010). Bilateral collaboration in water and wastewater services in Finland. *Water*, **2**(4), 815–825, www.mdpi.com/journal/water/ (accessed 22 March 2025), https://doi.org/10.3390/w2040815

Kurki V., Pietilä P. and Katko T. (2016). Assessing regional cooperation in water services: finnish lessons compared with International findings. *Public Works Management & Policy*, **21**(4), 368–389, https://doi.org/10.1177/1087724X16629962

Kuulas A., Renko T. and Kuivamäki R. (2020). (Original in Finnish) The investment requirements of water services infrastructure by 2040. FIWA Handout Series No. 63, Finnish Water Utilities Association (FIWA), Helsinki, Finland, https://www.vesilaitosyhdistys.fi/verkkokauppa/tuotteet/vesihuollon-investointitarpeet-vuoteen-2040/ (accessed 22 March 2025).

Ladner A., Keuffer N. and Baldersheim H. (2016). Measuring Local Autonomy in 39 Countries (1990–2014). *Regional & Federal Studies*, **26**(3), 321–357, https://doi.org/10.1080/13597566.2016.1214911

Lamberg J.-A., Ojala J. and Eloranta J. (1997). Uusinstitutionalismi ja taloushistoria. Kollektiivisen valinnan ja liiketoiminnan kustannusten problematiikka (New institutionalism and history of economics. Collective choice and the problematique of transaction costs). In: Uusi institutionaalinen taloushistoria. Johdanto tutkimukseen. (New Institutional Economics. Introduction to Research), J.-A. Lamberg and J. Ojala (eds), Atena Kustannus Oy, Jyväskylä, Finland, pp. 15–47.

Landriani L, Lepore L., D'Amore G., Pozzoli S. and Alvino F. (2019). Decorporatization of a municipal water utility: a case study from Italy. *Utilities Policy*, **57**, 43–47, https://doi.org/10.1016/j.jup.2019.01.005

Laurinolli H. (2011). Tilit kunnossa, kunta kuralla (Accounts in balance, municipalities not). *Aikalainen*, **16**(11), 4.

Leppälä K. (1998). Miten tekniikkaa oikein tieteellisesti tutkitaan? (How to study technology in scientific ways?) *Tiedepolitiikka*, **15**(2), 25–30.

Lieberherr E. (2011). Regionalization and water governance: a case study of a Swiss wastewater utility. *Procedia–Social and Behavioral Sciences*, **14**, 73–89, https://doi.org/10.1016/j.sbspro.2011.03.026

Lipford J. and Yandle B. (1997). Exploring the production of social order. *Constitutional Political Economy*, **8**, 37–55; cited by Mantzavinos et al., 2004, p. 79, https://doi.org/10.1023/A:1009037905120

Lobina E. (2013). Remediable institutional alignment and water service reform: Beyond rational choice. *International Journal of Water Governance*, **1**, 109–132, p. 121, https://doi.org/10.7564/12-IJWG3

Lobina E. (2018). Commentary on the European Commission's 'Study on Water Services in Selected Member States'. PSIRU, University of Greenwich, London, p. 6.

Lobina E., Weghmann V. and Nicke K. (2021). Water Remunicipalisation in Paris, France and Berlin, Germany. PSIRU, University of Greenwich, London.

Mahajan R. R. and Rajankar M. (2024). Inclusive management of our water commons. *Ecology, Economy, and Society– the INSEE Journal*, **7**(1), 29–46, https://doi.org/10.37773/ees.v7i1.1161

Mannermaa M. (1991). Evolutionaarinen tulevaisuudentutkimus (Evolutionary Futures Research). Acta Futura Fennica no. 2. Painatuskeskus, Helsinki, Finland.

Mannermaa M. (1993). (Original in Finnish) Soft systems methodology in futures studies. Pehmeä systeemimetodologia tulevaisuudentutkimuksessa, pp. 89–95. In: (Original in Finnish) Miten tutkimme tulevaisuutta? Kommunikatiivinen tulevaisuudentutkimus Suomessa. How do we Explore our Futures? The Communicative Futures Research in Finland, M. Vapaavuori (ed.), The Finnish Society for Futures Studies, Helsinki, Finland, p. 94.

Mantzavinos C., North D. C. and Shariq S. (2004). Learning, institutions, and economic performance. *Perspectives on Politics*, **2**(1), 75–84, p. 79, https://doi.org/10.1017/S1537592704000635

Marin P. (2009). Public-Private Partnerships for Urban Water Utilities: a Review of Experiences in Developing Countries. Trends and Policy Options No. 8, The International Bank for Reconstruction and Development/The World Bank, Washington, DC, p. 13.

Marin P., Fall M. and Ouibiga H. (2010). Corporatizing a water utility. A successful case using a performance-based service contract for ONEA in Burkina Faso. Gridlines, no. 53, https://openknowledge.worldbank.org/entities/publication/6ed5ca9a-6c44-5b08-9aab-79774061c288 (accessed 22 March 2025).

Marjoram T. (2010). Engineering: Issues, Challenges and Opportunities for Development (UNESCO report). UNESCO, Paris, France, https://unesdoc.unesco.org/images/0018/001897/189753e.pdf#212244 (accessed 22 March 2025).

Mashauri D. A. and Katko T. S. (1993). Water supply development and tariffs in Tanzania: from free water policy towards cost recovery. *Environmental Management*, **17**(1), 31–39, https://doi.org/10.1007/BF02393792

Mattila H. (2005). Appropriate management of on-site sanitation. Doctoral dissertation no. 537, Tampere University of Technology, Tampere, Finland, https://trepo.tuni. fi//handle/10024/114828 (accessed 22 March 2025).

Mattisson O. (1993). Va i England, Frankrike och Tyskland (Water and sewerage in England, France and Germany). *VAV-nytt*, **2**, 13–15.

McCaffrey S. (1992). A human right to water: domestic and international implications. *Georgetown International Environmental Law Review*, **5**, https://www.scirp.org/ reference/referencespapers?referenceid=3274131 (accessed 22 March 2025).

McDonald D. (2022). Meanings of Public and the Future of Public Services. Routledge. London, UK. DOI: 10.4324/9781003293002

McDonald D.A. (2016). To corporatize or not to corporatize (and if so, how?). *Utilities Policy*, **40**, June, 107–114. http://dx.doi.org/10.1016/j.jup.2016.01.002

Melosi M. V. (2005). Path dependency and urban history: is a marriage possible? In: Resources of the City, D. Schott, B. Luckin and G. Massard-Guilbaud (eds), Chapter 16, Routledge, London, UK.

Melosi M. V. (2008). The sanitary city. Environmental Services in Urban America from Colonial Times to the Present. Abridged edition. University of Pittsburg Press, Pittsburg, Pennsylvania, USA.

Mäntysalo R., Kallio O., Niemi P., Vakkuri J. and Tammi J. (2017). Finnish local government reform: juxtaposing cost structures and the centre periphery relations of municipalities in urban regions. In: Nordic Experiences of Sustainable Planning: Policy and Practice, S. Kristjánsdóttir (ed.), Routledge, London, pp. 243–262.

Nasdaq. (2021). United Utilities to Sell Stake in Tallinn Water – Quick Facts. Feb 3, 2021, https://www.nasdaq.com/articles/united-utilities-to-sell-stake-in-tallinn-water- quick-facts-2021-02-03 (accessed 22 March 2025).

Nevala A., Häyrynen S. and Korpela J. (eds) (2023). Fantastinen yliopistouudistus (Fantastic university reform). Sopeutuva akateemisuus ja sivistysihanteen muutos. Kirjokansi, Joensuu, Finland.

Newman P. and Thorley A. (1996). Urban Planning in Europe. International Competition, National Systems and Planning Projects. Routledge, London and New York.

North D. C. (1990). Institutions, Institutional Change and Economic Performance. Cambridge University Press, USA, p. 4.

North D. C. (1993). The New Institutional Economics and Development. EconPapers, Economic History, 9309002. University Library of Munich, Germany.

North D. C. (1994). The historical evolution of polities. *International Review of Law and Economics*, **14**(4), 381–391, https://doi.org/10.1016/0144-8188(94)90022-1

North D. C. (2003). The Role of Institutions in Economic Development. UNECE Discussion Papers Series No. 2003.2. Geneva, Switzerland, p. 2, https://unece.org/ fileadmin/DAM/oes/disc_papers/ECE_DP_2003-2.pdf (accessed 22 March 2025).

Oakerson R. J. (1999). Governing Local Public Economies: Creating the Civic Metropolis. ICS Press, Oakland, CA, USA.

OECD (Organisation for Economic Co-Operation and Development). (2009). Managing Water for All: An OECD Perspective on Pricing and Financing, p. 86, https:// www.oecd-ilibrary.org/environment/managing-water-for-all_9789264059498-en (accessed 22 March 2025).

OECD (Organisation for Economic Co-Operation and Development). (2010). Pricing Water Resources and Water and Sanitation Services. OECD Publishing, Paris, p. 18, https://www.oecd-ilibrary.org/environment/pricing-water-resources-and- water-and-sanitation-services_9789264083608-en (accessed 22 March 2025).

OECD. (2014). Water Governance in the Netherlands: Fit for the Future? OECD Studies on Water, OECD Publishing, Paris, France, https://dx.doi. org/10.1787/9789264102637-en.

OECD (Organisation for Economic Co-Operation and Development). (2020). Financing Water Supply, Sanitation and Flood Protection: Challenges in EU Member States and Policy Options. OECD Studies on Water. OECD Publishing, Paris, France, https://www.oecd.org/en/publications/financing-water-supply-sanitation-and-flood-protection_6893cdac-en.html (accessed 22 March 2025).

OECD/UN and ECA/AfDB. (2022). Africa's Urbanisation Dynamics 2022: The Economic Power of Africa's Cities, West African Studies. OECD Publishing, Paris, France, p. 128–129, https://doi.org/10.17

Ofwat (The Water Services Regulation Authority). (2024a). Monitoring Financial Resilience Report 2023–24, p. 4, https://www.ofwat.gov.uk/wp-content/uploads/2024/11/Monitoring-Financial-Resilience-Report-2023-24.pdf (accessed 22 March 2025).

Ofwat (The Water Services Regulation Authority). (2024b). Water company performance report 2023–24, p. 9, https://www.ofwat.gov.uk/regulated-companies/company-obligations/outcomes/water-company-performance-report-2023-24/ (accessed 22 March 2025).

OHCHR, UN-HABITAT and WHO. (2010). The right to water. Fact Sheet No. 35, https://www.ohchr.org/en/publications/fact-sheets/fact-sheet-no-35-right-water (accessed 22 March 2025).

Ostrom E. (2005). Understanding Institutional Diversity. Princeton University Press, Princeton, New Jersey, USA. https://doi.org/10.1515/9781400831739

Ostrom E. (2010). Beyond markets and states: polycentric governance of complex economic systems. *American Economic Review*, **100**, 641–672; p. 653, https://doi.org/10.1257/aer.100.3.641

Ostrom V. and Ostrom E. (1991). Public goods and public choices: the emergence of public economies and industry structures. In: The Meaning of American Federalism, V. Ostrom (ed.), Institute for Contemporary Studies Press, San Francisco, CA, USA, pp. 163–197.

Oulasvirta L. (1994). Uusi organisaatioiden taloustiede ja kunnallishallinnon tutkimus. New organisational economics and research on municipal administration. In: A. V. Anttiroiko (ed.). Kunnallishallinto ja politiikan taloustiede. Uusi poliittinen taloustiede kuntien hallinnon, talouden ja ympäristösuhteiden analyysikehyksenä (Municipal Administration and the Economics of the Politics. New Political Economics as an Analytical Framework for Administration, Economy and Environmental Relations of Municipalities). KT-Tietokeskus, Tampere, Finland. pp. 146–161, p. 147.

Our World in Data. (2024). Elected local governments, https://ourworldindata.org/grapher/strong-elected-local-governments-index (accessed 22 March 2025).

Ozekin K. (2024). 15 Dec 2024, Personal communication, Water Research Foundation, Denver.

Pacey A. (1977). Technology is not enough: the provision and maintenance of appropriate water supplies. *Aqua*, **1**(1), 1–58.

Pacey A. (1983). The Culture of Technology. Blackwell, London.

Parviainen S. (2019). Informal practices in economy (Epäviralliset käytännöt taloudessa). *The Finnish Economic Journal*, **115**(3), 552–557, pp. 555–556.

Peda P. and Vinnari E. (2020). The discursive legitimation of profit in public-private service delivery. *Critical Perspectives on Accounting*, **69**, 102088, https://doi.org/10.1016/j.cpa.2019.06.002

Peltomaa J. (2013). Vesihuollon moninaiset merkitykset (Diverse meanings of water services). *Alue ja Ympäristö*, **42**(2), 102–104.

Pérard E. (2009). Water supply: Public or private? An approach based on cost of funds, transaction costs, efficiency and political costs. *Policy and Society*, **27**, 193–219, https://doi.org/10.1016/j.polsoc.2008.10.004

Pietilä P. (2013). Diversity of the water supply and sanitation sector: roles of municipalities in Europe, pp. 99–111. In: Water Services Management and Governance: Past Lessons for a Sustainable Future, T. Katko, P. Juuti and K. Schwartz (eds), IWA Publishing, London, UK. Open access, https://iwaponline.com/ebooks/book/480/Water-Services-Management-and-Governance (accessed 22 March 2025).

Pietilä P., Katko T. and Kurki V. (2010). Vesi kuntayhteistyön voiteluaineena (Water fueling municipal collaboration). Publ no 62, Foundation for Municipal Development, Helsinki, Finland, p. 95, https://kaks.fi/wp-content/uploads/2010/10/Pietil%C3%A4_Katko_Kurki.pdf (accessed 22 March 2025).

Piironen O. (2013). The Transnational Idea of University Autonomy and the Reform of the Finnish Universities Act. *High Educ Policy*, **26**, 127–146, https://doi.org/10.1057/hep.2012.22

Plimmer G. and Pickard J. (2018). Gove attacks water industry over high pay and dividends. *Financial Times*, 1 March 2018.

Prasad N. (2006). Privatisation results: private sector participation in water services after 15 years. *Development Policy Review*, **24**(6), 669–692, p. 688, https://doi.org/10.1111/j.1467-7679.2006.00353.x

Price W. (2001). A Pracademic Research Agenda for Public Infrastructure. *Public Works Management & Policy*, **5**, 287–296, https://doi.org/10.1177/1087724X0154004

PSIRU. (2014). Public and private sector efficiency. A briefing for the EPSU Congress by PSIRU.

Public Futures. (2024). The database of remunicipalisations, https://publicfutures.org/cases (accessed 22 March 2025).

Rajala R. P., Katko T. S. and Springe G. (2019). Students' perceived priorities on water as a human right, natural resource, and multiple goods. *Sustainability*, **11**(22), 6354, https://doi.org/10.3390/su11226354

Rasenberg E. (2021). EC: Member States should raise water prices. Water News Europe. 27 April 2021, https://www.waternewseurope.com/member-states-should-raise-water-prices/ (accessed 22 March 2025).

Raunio T. and Saari J. (eds) (2006). Eurooppalaistuminen. Suomen sopeutuminen Euroopan integraatioon (Europeanisation. Finland's Adaptation to European Integration). Gaudeamus, Helsinki, Finland, pp. 234–235.

RoK (Republic of Kenya). (2016). Kenya Gazette Supplement No. 164 (Acts No.43). Nairobi, 20 Sept 2016.

ROTI. (2017). Summary of recommendations. Water Supply and Sewage Services in Need of a Structural Update, https://www.ril.fi/en/alan-kehitys/the-roti-2017-report/summary/summary-of-recommendations.html (accessed 15 Dec 2024).

Ryynänen A. (2012). Kunnallishallinnon kansainväliset vaikutteet (International Influences of Municipal Administration). Tampere University Press, Tampere, Finland.

Sandelin S., Hukka J. and Katko T. (2019). Importance of Knowledge Management at Water Utilities. *Public Works Management & Policy*, **26**, https://doi.org/10.1177/1087724X19870813

Schumacher E.F. (1973). Small is Beautiful. Economies as if People Mattered. Harper/Perennial, New York, USA.

Scott W. R. (2008). Institutions and Organizations. Ideas and Interests. Sage Publications, USA; cited by Hatakka A. (2016).

Seppälä O. (2019). Pidetään yhteiskunnan verisuonisto kunnossa (Let us keep the blood systems of society in good condition). Vieras. *Rakennustekniikka*, **75**(2), 21.

Seppälä O. T. and Katko T. S. (2003). Appropriate pricing and cost recovery in water services. *Journal of Water Supply: Research and Technology – AQUA*, **52**, 225–236, https://doi.org/10.2166/aqua.2003.0022

Thackston E. L., Parker F. L., Minor M. S., Bowen J. D. and Godwin W. S. (1983). Water Policy in Tennessee: Issues and Alternatives. Vanderbilt University, Nashville, Tennessee, USA.

UCLG (United Cities and Local Governments) Policy Paper. (2016). The Role of Local Governments in Territorial Economic Development. UCLG Congress, Bogota, Columbia, p. 28, https://www.uclg.org/sites/default/files/the_role_of_local_governments_in_territorial_economic_development.pdf (accessed 22 March 2025).

UN (United Nations). (2016). Water and jobs. UN Word Water Development Report, p. 3, https://www.unesco.org/en/wwap/wwdr/2016 (accessed 15 Dec 2024).

UN (United Nations). (2021). The United Nations World Water Development Report 2021: Valuing Water. UNESCO, Paris, France, p. 10, https://www.unwater.org/publications/un-world-water-development-report-2021 (accessed 22 March 2025).

United Nations Economic Commission for Europe (UNECE) and World Health Organization (WHO) Regional Office for Europe. (2019). The Human Rights to Water and Sanitation in Practice: Findings and lessons learned from the work on equitable access to water and sanitation under the Protocol on Water and Health in the pan-European region, p. 11, https://unece.org/environment-policy/publications/human-rights-water-and-sanitation-practice-findings-and-lessons (accessed 22 March 2025).

UNEP. (2012). Measuring water use in a green economy. In: A Report of the Working Group on Water Efficiency to the International Resource Panel, J. McGlade, B. Werner, M. Young, M. Matlock, D. Jefferies, G. Sonnemann, M. Aldaya, S. Pfister, M. Berger, C. Farell, K. Hyde, M. Wackernagel, A. Hoekstra, R. Mathews, J. Liu, E. Ercin, J. L. Weber, A. Alfieri, R. Martinez-Lagunes, B. Edens, P. Schulte, S. von Wirén-Lehr and D. Gee (eds), https://www.resourcepanel.org/reports/measuring-water-use-green-economy (accessed 22 March 2025).

United Nations Conference on Environment & Development (UNCED). (1992). Agenda 21. Rio de Janeiro, Brazil, 3 to 14 June 1992, https://sustainabledevelopment.un.org/outcomedocuments/agenda21 (accessed 22 March 2025).

UN-Water Decade Programme on Advocacy and Communication (UNW-DPAC). (2011). The Human Right to Water and Sanitation: Milestones, https://www.un.org/waterforlifedecade/pdf/human_right_to_water_and_sanitation_milestones.pdf (accessed 15 Dec 2024).

UN-Water Decade Programme on Advocacy and Communication (UNW-DPAC) and Water Supply and Sanitation Collaborative Council (WSSCC). (2011). The human right to water and sanitation. Media Brief, https://www.un.org/waterforlifedecade/pdf/human_right_to_water_and_sanitation_media_brief.pdf (accessed 15 Dec 2024).

Vakkuri J. (2010). 'Suureksi kasvamisen dilemma– suuruuden ekonomia kuntakehittämisen periaatteena (The dilemma of growing large– the economy of size as a principle of municipal development). *Kunnallistieteellinen aikakauskirja*, **38**(2), 117–119.

Verwoerd J. (2024). The Paradox of Decentralization: Impact on Corruption in Water Utilities, https://www.preprints.org/manuscript/202409.1652/v1 (accessed 22 March 2025).

Vinnari E. (2008). Public service or public investment? An assessment of the consequences of new public management in the water sector. Doctoral dissertation no. 726, Tampere University of Technology, Tampere, Finland, https://trepo.tuni.fi//handle/10024/114736 (accessed 22 March 2025).

Vinnari E. M. and Hukka J. J. (2007). Great expectations, tiny benefits – decision making in the privatization of Tallinn Water. *Utilities Policy*, **15**(2), 78–85, https://doi.org/10.1016/j.jup.2007.01.001

Vinnari E., Hukka J. J. and Katko T. S. (2005). Kunnallisen vesihuollon omistusjärjestelyt: Yhteisen edun ajamista vai lyhytnäköistä opportunismia? (Ownership arrangements of municipal water services: Seeking for public interest or short-sighted opportunism?) *Kuntatekniikka*, **60**(1), 20–21.

WASH (Water and Sanitation for Health). (1992). Opening the door to the private sector. Lesson learned forum. Information for action from the Water and Sanitation for Health Project. Arlington, Virginia, https://www.ircwash.org/sites/default/files/WASH-1993-Lessons.pdf (accessed 22 March 2025).

WASREB (Water Services Regulatory Board). (2018). Guideline on Clustering of Water Service Providers, https://wasreb.go.ke/wp-content/uploads/2023/08/Guideline-on-Clustering-2018.pdf (accessed 22 March 2025).

WASREB. (2021). Impact: A Performance Report of Kenya's Water Services Sector – 2019/20. Water Services Regulatory Board, Nairobi, https://wasreb.go.ke/impact-reports/ (accessed 22 March 2025).

WASREB (Water Services Regulatory Board). (2024). A Performance Report of Kenya's Water Services Sector – 2022/23. Issue No. 16–2024, pp. 50–51, https://wasreb.go.ke/wp-content/uploads/2024/06/Impact-Report-16.pdf (accessed 22 March 2025).

WaterFront. (2022). World could fall short of 10 million water professionals. *WaterFront Daily*. 24 Aug 2022, https://www.worldwaterweek.org/news/world-could-fall-short-of-10-million-water-professionals (accessed 22 March 2025).

WWAP (World Water Assessment Programme). (2012). The United Nations World Water Development Report 4, Vol. 1: Managing Water under Uncertainty and Risk. UNESCO, Paris, France, https://sustainabledevelopment.un.org/content/documents/404water.pdf (accessed 22 March 2025).

Yearwood K. (2018). The Privatised Water Industry in the UK. An ATM for investors. PSIRU, University of Greenwich, London.

Zajda J. (2006). Introduction: Decentralisation and Privatisation in Education: The Role of the State. In: Decentralisation and Privatisation in Education, J. Zajda (ed.), Springer, Dordrecht, the Netherlands, https://doi.org/10.1007/978-1-4020-3358-2_1

doi: 10.2166/9781789064162_0141

Chapter 5

The future of water services

'Never measure the height of a mountain until you have reached the top. Then you will see how low it was'. (Dag Hammarskjold)

'We have the means and the capacity to deal with our problems, if only we can find the political will'. (Kofi Annan)

5.1 INTRODUCTION

Water services, like societies as a whole, are subject to an ever-changing operational environment, which ultimately sets requirements that we need to be prepared for in advance so we can adapt to these new conditions. The Office of the Director of National Intelligence (2021) stated that governments, industry and civil society will face an increasing risk of water insecurity over the next two decades as demand grows and supply is increasingly strained. Moreover, poor governance and resource management, development practices, agriculture, and environmental degradation are also likely to diminish the quantity and quality of water supplies in many parts of the world, unless clear improvements are made.

Climate change, demographic and economic transformation, and technological and behavioural change are complex processes and their interrelations will affect both the availability of water resources and the way they are used and shared. Therefore, public water utilities will continue to have a fundamental responsibility in finding effective solutions that ensure a more sustainable and equitable access to water resources for all (European Association of Public Water Operators, 2019). In self-supply water and on-site sanitation, the owner of real estate is in charge while services can be bought from local enterprises and hopefully adequate support be gained from local authorities.

Future trends in water demand are difficult to predict accurately. Burek *et al.* (2016) estimated that overall, global demand for water will continue to increase at an annual rate of about 1%, resulting in an increase of between 20 to 30% by 2050, with a margin of error of more than 50%. The evolution of water demand is highly location specific, reflecting shifting use patterns across the three major water use sectors – municipalities, industries and agriculture.

The 2030 Agenda for SDG6 aims to 'ensure availability and sustainable management of water and sanitation for all' and includes targets for universal access to safe drinking water (6.1), and sanitation and hygiene (6.2). Data for the corresponding global indicators are now available for more than half of the world's population although the world is not on track to achieve these targets as the mid-point of the SDG period is reached (UNICEF and WHO, 2023, p. vi).

In 2022, as many as 2.2 billion people lacked safely managed drinking water, including 1.5 billion with basic services, 292 million with limited services, 296 million with unimproved services and 115 million people having access only to surface waters (UNICEF and WHO, 2023, p. 158). Furthermore, 3.5 billion people lacked safely managed sanitation, including 1.9 billion with basic services, 570 million with limited services, 545 million with unimproved services, and 419 million practising open defecation. Achieving universal access to safely managed water services by 2030 requires a six-fold increase in current rates of progress (this is 20-fold in least developed countries and 19-fold in fragile contexts). Achieving universal access to safely managed sanitation would require a fivefold increase in current rates of progress (16-fold in least developed countries and 15-fold in fragile contexts).

In reality, some of those who have access to water supply and sanitation (WSS) services, however, often have only intermittent water supply, sewerage system overflows, and poor customer service (Mumssen *et al.*, 2018 cited by Lombana Cordoba *et al.*, 2022, p. ix). The absence of safely managed water services and drinking water and sanitation facilities causes vast economic losses to citizens and businesses, environmental degradation, social inequality, and takes a huge toll on dignity, health and wellbeing in communities worldwide. The ageing and decaying water infrastructure – a creeping crisis – exacerbates these problems in non-OECD countries and comes at a large financial cost including a sizeable loss of economic activity, as is also the case in OECD countries. Therefore, it would be incorrect to consider this unsustainability challenge as being limited to the developing world; rather, it is a global challenge that also concerns everyone in the developed world (Hukka & Katko, 2021, p. 59).

According to the International Law Association (2004, p. 22), however, in determining the need for future equitable and reasonable use of water resources in the basin nations, they shall first allocate waters to *satisfy vital human needs*. UNESCO (2023) pointed out that the current inadequate rate of progress towards the SDG6 targets highlights the need to explore opportunities through partnerships and cooperation. Cooperation improves water governance and decision making, stimulates innovative solutions and leverages efficiencies. By promoting *inclusive engagement*, participation and dialogue, and giving voices to those who are otherwise unheard, partnerships can help ensure that no one is left behind and that human rights to water and sanitation are realised.

The growing recognition by infrastructure organisations of the need to improve effectiveness, overall operational performance and reliability requires *clear understanding* of how to manage infrastructure assets in a way that allow their current performance to improve and be competitive, while ensuring that they are planning and reinvesting for the future. Consequently, organisations that manage infrastructure assets are driven to adopt a formal and holistic approach in order to provide services in the most cost-effective way and to demonstrate this to customers and stakeholders (Too, 2012).

In light of an unpredictable and rapidly changing operating environment, the World Bank has developed a programme to guide utilities in initiating and maintaining transformation efforts. The goal is to become a Utility of the Future (UoF) – a future-focused utility that provides high-quality, reliable, safe, inclusive, transparent and responsive WSS services through best-fit practices in an efficient, resilient, and sustainable manner. The UoF is a new paradigm for providing innovative, inclusive, and market and customer oriented services far beyond what most utilities have yet achieved or have even aimed at. At the lower level of the success pyramid sits the legal framework and governance, the next level includes financial and human resources management, organisation and strategy. The third level involves commercial management and technical operations, with service to customers at the top of the pyramid. (Lombana Cordoba *et al.*, 2022, p. xi, adopted from Heymans *et al.* 2016).

There are several definitions of viable water utilities and resilience formulated by various well-known organisations and prominent scholars. Based on those definitions, Hukka and Katko (2021, p. 73) have formulated state-of-the-art definitions for a viable water utility (Box 5.1) and for the resilience of the water utility of the future (Box 5.2):

Box 5.1 A viable water utility

A viable water utility is a well-managed, service-oriented, customer-friendly, resilient and proactive public or user-owned entity. It has managerial and technical competence as well as financial capability to produce and maintain safe, reliable and preferred services, which meet regulatory and performance requirements as well as its own economic, social, environmental and business objectives by minimizing the total cost of owning and operating infrastructure capital assets, now and in the future. A viable water utility promotes and actively participates in the development of regulatory governance, provision, and production to enhance sustainability of water services. It also has reciprocal and responsive relationship with other utilities to facilitate resilience.

Source: Hukka and Katko (2021, p. 73).

> **Box 5.2 Resilience of a water utility**
> The competence of a utility – also by extending and readjusting its response capabilities and resources needed – to predict, prepare for, adapt to, withstand, communicate and recover promptly, efficiently, and effectively from the consequences of any human-caused intentional and unintentional or naturally occurring hazard, threat or incident – both foreseeable and unexpected – in order to successively produce and maintain safe, reliable and preferred services, to protect the society and the environment, and to learn from the experience gained, now and in the future.
> Source: Hukka and Katko (2021, p. 73).

The North Carolina State Water Infrastructure Authority (2017) recognised that best practices in utility management are essential for viable utility systems. These practices are grouped into three integrated focus areas (Figure 5.1).

In decision making and management, we normally recognise three timeframes. In operative or operational management, the timeframe is up to one year, in strategic management up to 10 years, and in visionary management up to 30 or even 50 years (Myth 18, Kaivo-oja *et al.*, 2004, p. 530). Furthermore, Kaivo-oja *et al.* (2004, p. 534) noted that management is essentially related to decision making. Within an organisation, decision making is constrained by the principle of bounded rationality and path dependence. North (2003) argued that the institutions and beliefs of the past have an enormous effect on constraining the ability to make change both in the present and future.

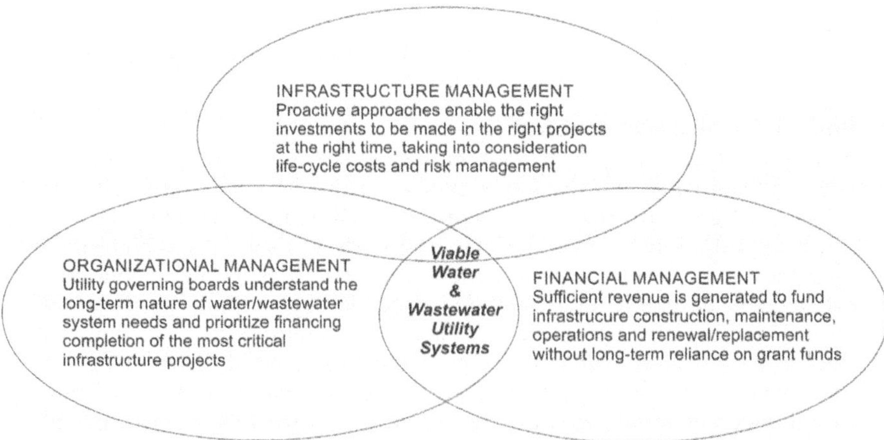

Figure 5.1 The best practices in utility management for viable utility systems.
Source: North Carolina State Water Infrastructure Authority (2017).

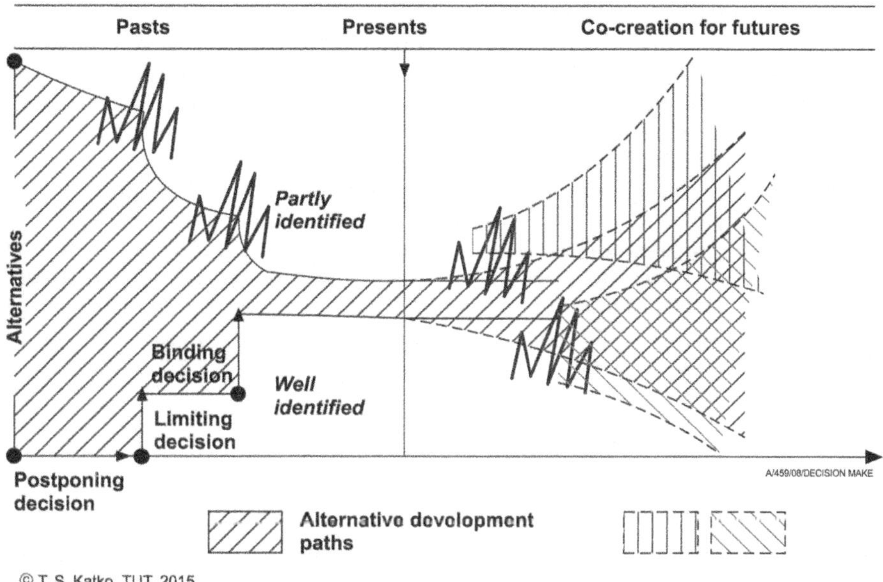

© T. S. Katko, TUT, 2015

Figure 5.2 Path dependence and related decision making.
Source: Katko (2016, p. 47).

Figure 5.2 illustrates a schematic proposal for how certain strategic decisions can be linked to path dependence and how they may affect available future development options or paths. By their nature, decisions can be classified as binding, limiting or postponing. The combination of historical research (HR) and futures research (FR) offers probably the best opportunity to analyse alternative and co-created futures. The key argument is that sustainable and viable development unavoidably requires clear understanding of the pasts and presents as a basis for formulating potential, probable and desirable futures (Antila *et al.*, 2013; Kaivo-oja *et al.*, 2004). In the context of water systems, the planning period can often be about 30 or even 50 years, while the lifetime of the infrastructure may exceed 100 years, as can networks (Myth 18: Kaivo-oja *et al.*, 2004, p. 535; Katko 2016, p. 47).

In most countries, funding for water infrastructure is neither adequate nor sustainable. Water services remain low on the scale of policy priorities despite well-documented contributions to human and economic development (UN-Water, 2015). Therefore, the challenge of meeting future demands and achieving sustainable water and sanitation for all is significant. Strong political will and commitment are required. However, as UN-Water (2018) emphasised, there is no standard approach for sequencing the policies, institutional developments and investments required for effective management of water resources, and the provision and production of services that would be valid for all countries and under all circumstances.

For example, in New Zealand, the Office of the Auditor-General (2014) recommended that decisions on how to manage infrastructure need to be made in the context of each local authority's financial position and prospects, and its community circumstances. Local authorities cannot separate decisions on their assets and service delivery from considerations about funding sources and timings of funding and renewal or replacement work. Asset management decisions are so closely aligned with a local authority's financial strategy that any weak link can undermine the strength of the asset management chain.

In addition, professional will is needed, i.e. willingness to view the development of water services in a wider institutional context. Broader collaboration with stakeholders of water services is also required, which is a challenge to sector professionals and stakeholders as highlighted by Katko (2016, pp. 243–244).

Myth 18: Future problems can be solved without considering history

Figure 5.3 A commonly used argument: we are not interested in the past, only the future. Is this possible?
Illustration: Pertti Väyrynen.

Since ancient times, the need for fresh, safe and healthy water has resulted in the development of various kinds of water supply and sanitation systems, water purification devices and treatment methods (e.g. Juuti *et al.*, 2007). The necessity for fresh water has influenced individual lives as well as whole societies.

Juuti (2001) distinguished the following five development phases of water and sanitation services. First, preurban systems prior to the establishment of cities (c. 10 000 A.D. and later); second, early systems up until 500 A.D.; and third, the stagnation period up to 1700 A.D. The fourth period of slow development in the Finnish case continued until 1910, followed by modern urban infrastructure. Most point source wastewaters were properly treated by the 1980s.

Juuti (2001, p. 270) classified water supply and sanitation systems according to three development phases: bucket systems, prototypes and modern systems. In Finland, bucket systems were gradually changed to prototypes including among other options, low pressure water networks without treatment, and water toilets and combined sewers that discharged wastewaters to water bodies without treatment. Modern systems that included groundwater use and wastewater treatment and later separate sewers were also introduced gradually from 1910 and in particular, after the second world war.

Sedlak (2014) defined water systems over the past 2500 years into three major phases. Water 1.0 included the early Roman aqueducts, fountains and sewers that made dense urban living feasible. He considers the development of drinking water as Water 2.0 and sewage treatment systems as Water 3.0. We are now in phase Water 4.0 that includes reduced water use, conservation and options for centralised or decentralised, and even household level systems (Sedlak, 2024). Sedlak contends that modern water infrastructure still adheres to the original blueprint for ancient Rome. Sedlak (2021) outlined the three major themes of the 'Next Path': (i) the IT revolution, (ii) brackish water and treated effluent as sources of water, and (iii) combining IT and modular water treatment technology for safe, small-scale treatment and supply networks.

In his recent book, Gleick (2023) distinguished three major eras of water management. The First Age of Water starts with the Big Bang continuing through early human history, including the first agricultural developments along major rivers, the earliest empires that constructed dams and aqueducts, the birth of the first water institutions and water laws, and the first wars over the control of water. In the Second Age of Water, our current era, we have discovered the causes of water-related disease and water pollution. These solutions include water infrastructure to clean waters, to produce hydroelectricity and protect floods and droughts. We developed scientific tools for understanding the hydrologic cycle and its role in relation to climate and ecosystems, and the instruments with which to search our own solar system and the distant universe for signs of water, and perhaps new forms of life beyond our own.

According to Gleick (2024), it is time for a more sustainable and positive future that is not only possible but inevitable. We can provide safe water and sanitation for all, reduce water poverty and disease, and grow more food more efficiently. We can also reduce violence over water resources through negotiation and cooperation. Ultimately, we can stop climate change and reduce the threats to water resources and humanity.

Occasionally, there are claims – even by water service professionals – that we are not interested in history but only in the future, as if these were alternatives or were not interdependent (Figure 5.3). Traditionally, historical research seldom extends to the present day. Similarly, futures research may ignore the limitations that history and earlier decisions have made relating to the present day and thus to the futures. If these research fields are too strongly tied to their own traditions, they both lose wider vision and do not perceive the entire potential scope (Kaivo-oja *et al.*, 2004, pp. 540–543; Katko *et al.*, 2009).

According to Ruonavaara (2006), social phenomena derive their development history from various events and stages. When an alternative is chosen, it is not easy to deviate from the selected path. The same line is continued until a situation is reached where new, alternative choices appear.

Nevertheless, past events and decisions inevitably impact the present and the future. Decisions made should be known and understood at least on a general level. Although the past does not necessarily repeat itself or take the same form, it is still worthwhile to keep in mind the words of the American philosopher George Santayana in 1905: 'Those who cannot remember the past, are condemned to repeat it.' (Santayana, 1905).

REALITY: Many repeat the mistakes of our historical past. If we clearly understand where we are coming from and where we are now, we can take active steps to identify and influence the future and development paths that are preferred.

In their book 'The Narrow Corridor', Acemoglu and Robinson (2019) explored how societies can maintain a balance between state power and the rights of citizens, avoiding the dangers of either tyranny or anarchy through inclusive institutions. Based on a study of the past one thousand years of human development, Acemoglu and Johnson (2023, cited by Naughton, 2023), the 2024 Nobel Laureates in Economics, argued that 'the broad-based prosperity of the past was not the result of any automatic, guaranteed gains of technological progress... Most people around the globe today are better off than our ancestors because citizens and workers in earlier industrial societies organised, challenged elite-dominated choices about technology and work conditions, and forced ways of sharing the gains from technical improvements more equitably'.

Another Nobel Laureate, Stiglitz (2023), contended that with the right political reforms, democracies can become more inclusive, more responsive to citizens, and less responsive to corporations and wealthy individuals. However, salvaging democratic societies would also require far-reaching economic reforms.

In water services and management, key technology-related choices include the used pipe materials (Myth 10) and raw water sources (Myth 3). Decisions made in regional and intermunicipal cooperation can also have an impact far into the future. Hopefully, these decisions will make sense and contribute to sustainable development.

Although some choices made can be abandoned or replaced with new solutions, water services networks and systems in communities cannot be changed without some restrictions; however, a new district water supply or a central wastewater treatment plant can significantly change existing systems.

The ideology of free water (Myth 14) and that of privatisation of water services (Myth 17), especially in developing and transition economies, are both examples of how history has an impact. Attempts to privatise water supply were unsuccessful in the West as early as the nineteenth century (Katko *et al.*, 2002). Water use in the Nile River has been debated for thousands of years, most recently over the Grand Ethiopian Renaissance Dam under completion in Ethiopia (Bezahler, 2024).

5.2 TIME SPAN OF WATER SERVICES

The futures (the plural form is used intentionally here by the authors) can be viewed through three timeframes: from the perspective of day-to-day operational activities (one year), from a strategic management perspective (10 years), and taking visionary management viewpoint (50 years) (Figure 5.4). In visionary thinking, the industry or organisation seeks to perceive a future state to which access is desirable and preferable. Seen through this perspective, efforts can be made to identify alternative development paths and strategies to implement this vision (Kaivo-oja *et al.*, 2004, p. 540).

In water services, the historical development of systems needs to be assessed over a lengthy timespan of between 100 to 250 years and a similar period for the futures (Katko, 2016, p. 250). Thus, instead of one year, the 'quarter of water services' should be counted from a millennium, a fundamentally different

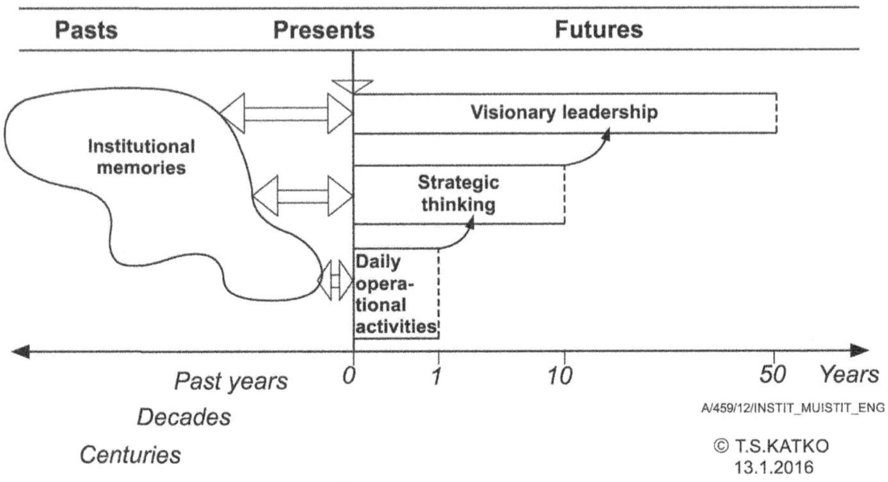

Figure 5.4 Timeframe for futures thinking and leadership: in addition to operational actions, we should have time for strategic and even visionary considerations.
Source: Katko (2016, p. 253).

timeframe. This is not at odds with the fact that daily operations should be managed in the best way possible. Besides, water utility managers and others hopefully will have time for strategic and visionary thinking (Katko, 2016, p. 257; Seppälä, 2004).

There are, however, different interpretations of the past as well as estimates of desired futures. The present is here and now, although it can be interpreted in many ways. Figure 5.5 elaborates on the links between pasts, presents and futures. All of these timespans can and should be used in decision making. Futures research can also take advantage of institutional memories. The present is bound by existing legislation and regulations, policy objectives and decisions that also affect futures. Analogues, i.e. conclusions and path dependencies

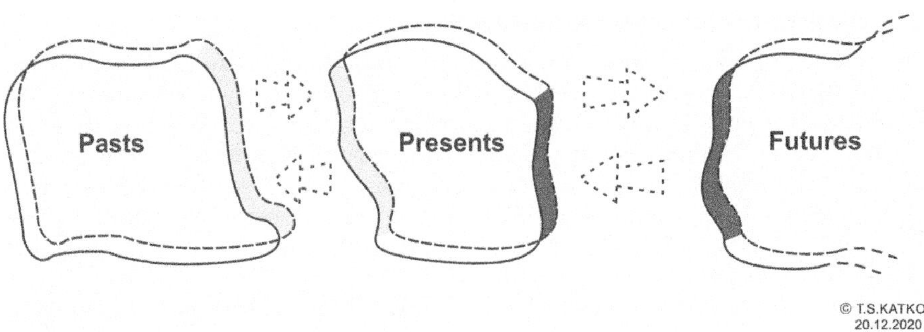

Figure 5.5 The interconnections of pasts, presents and futures.
Source: Katko *et al.* (2022, p. 166, modified).

based on similarity, connect these pasts, presents and futures. Therefore, pasts and presents are like newly separated continents, while presents and futures may already be further apart.

Institutional memories and tacit knowledge are important when producing alternative future goals and spaces. The basis of futures-thinking requires that tacit knowledge should be properly documented and analysed. This is relevant to historically constructed network materials and their location but also to long-term management traditions and habits (Sandelin, 2017).

Water services development rarely runs in a straight line or in one direction, as there are turning points and decisions that lead to new developments or one can revert to previous paths, such as choosing a raw water source. There will be constant surprises due to the changing operating environment. By linking history and the future, perspectives can also be expanded upon.

Myth 19: The poor cannot afford water services

Figure 5.6 In peri-urban settlements of developing economies, rather than piped and pumped systems, water is often transported using different methods at a high price and through the efforts of the poor themselves.
Illustration: Pertti Väyrynen.

In discussing this myth, the authors first explore the benefits that can be gained through safe water and proper sanitation. It is well recognised that water is connected to most aspects of economic and sustainable development (UN, 2021). As discussed in Myth 14, the United Nations Human Rights Council (HRC) adopted a consensus in September 2010, affirming that access to water and sanitation is a human right. This resolution, however, failed to provide guidance on how to sufficiently monitor progress or how the capital, operations and maintenance costs of water and sanitation services infrastructure can be provided while maintaining affordable prices for the poor (WWAP, 2012).

Skotnes (2016) explained that society-critical functions are essential to ensure the functioning of a society and its economy. The disruption of essential services – such as water supply and wastewater systems – may result in substantial economic damage. These critical infrastructures are the backbone of our modern and interconnected economies (OECD, 2019, p. 13).

In this context, OECD classifies six critical sectors: information and communication technologies, energy, finance, health, transport and water (2019, p. 48). The Critical Five Members (Australia, Canada, New Zealand, UK, and the USA) proposed that essential, nationally significant infrastructure covers the following: the systems, assets, facilities and networks that provide essential services and are necessary for national and economic security, prosperity, health and safety of their respective nations (Critical 5, 2014, p. 2). By ensuring that the critical infrastructure is secure and resilient, governments can protect and increase the strength and vitality of their respective economies (Critical 5, 2014, p. 4).

Water is arguably more fundamental than any other resource – to life itself, supporting ecosystem services – and to every economy and society (UNEP, 2012). It contributes directly and indirectly to virtually all other ecosystem services, while water and sanitation services are also an economic sector in themselves. Water is needed for all biotic and economic production processes, and in most societies, water also has important recreational, cultural and spiritual values. This fundamental threefold value – sustaining life, economies and cultures – creates an enormous range of competing demands on water resources but also offers economic opportunities (UNEP, 2012).

Water is a crucial element in ensuring gender equality and eradicating discrimination. Access to safe drinking water and sanitation is most at risk for those who have been denied rights to adequate housing, education, work or social security (OHCHR, 2010). Ensuring access to water and sanitation as human rights means this is a legal entitlement rather than a commodity or charitable service. Universal access to sanitation is 'not only fundamental for human dignity and privacy but is one of the principal mechanisms for protecting the quality' of water resources (UNW-DPAC and WSSCC, 2011).

Failure to improve water resource management could diminish national growth rates by as much as 6% of GDP by 2050 (World Bank, 2016, cited by UN and the World Bank, 2018). Yet, inadequate water supply and sanitation continues to have the greatest economic consequences of all water-related risks. It also continues to pose the most harmful risk to people, with diarrhoeal diseases resulting in 1.4 million premature deaths in 2010. The total global economic losses associated with inadequate water supply and sanitation have been estimated by WHO at USD 260 billion annually (Sadoff *et al.*, 2015). Deaths from water-related diseases are, however, inadequately monitored and reported; the best estimates analysed by Gleick (2002) fall between 2 and 5 million deaths per year.

However, it is likely that the benefits of safe and reliable water and sanitation services are systematically underestimated due to a number of non-economic benefits that are difficult to quantify but that are of high value to the individuals concerned in terms of dignity, convenience, social status, cleanliness, health and overall wellbeing. Furthermore, safe and reliable water and sanitation services appear to be a key driver for economic growth, including investments by firms that are reliant on sustainable water and sanitation services both for their production processes and their workers (OECD 2011a).

The COVID-19 pandemic underlined the fundamental role that clean water plays in the wellbeing of communities worldwide; safe water supply and sanitation are key requirements in promoting public health services proactively and they cannot be taken for granted (Katko & Hukka, 2021).

Whittington (2015) asserted that there can be a wide range of cost-benefit outcomes depending on local conditions. Therefore, it is argued that global averages of benefit-cost ratios (BCR) offer little useful information about the attractiveness of water, sanitation and hygiene (WASH) investments in local settings. Hutton and Varughese (2016) estimated that achieving universal access to safe drinking water, sanitation and hygiene (SDG targets 6.1 and 6.2) in 140 low- and middle-income countries would cost approximately USD 1.7 trillion between 2016 to 2030. The BCR of such investments has been shown to provide a significant positive return in most regions (WHO, 2012a). An analysis by Hutton (2018) suggested that current BCRs favour drinking water supply (at 3.4 and 6.8 for urban and rural areas respectively) over sanitation (with 2.5 and 5.2 for urban and rural areas respectively).

A common belief is that poor people cannot afford water services and the required amount of potable water. Since there is no free water (Myth 14), the aim is that the price of potable water and sanitation should be reasonable. The prices and delivery systems should be appropriate considering users' ability to pay.

In peri-urban settlements of developing economies water is often transported (Figure 5.6) rather than developing services based on piped and pumped systems and appropriate rules of the game. The limited funds are barely sufficient to keep the existing systems in place and are not enough to provide water supply to those who do not have it. In this way, those who are able to pay receive water services, but the poorest are easily deprived of them. Slum dwellers often have to pay an exorbitant amount for water that is transported by trucks or other equipment. It would be desirable for solvent customers to pay an appropriate price for water so that the systems can be properly run and developed. Transporting water is simply too expensive compared the cost of pumping it.

REALITY: In low-income economies, the poorest people pay the most for their water relative to their income. Rather than providing water for free or privatising water services, communities should consider improving their water services on the basis of public or joint ownership or water cooperatives where applicable.

An even greater challenge is sanitation. In 2000, altogether 1.3 billion people globally practised open defecation (Kashiwase, 2019); by 2022, this had declined to 420 million, with sub-Saharan Africa accounting for the largest share, followed by Southern Asia (Kashiwase, 2023). In 2023, estimated 1.7 billion people still lacked access to basic sanitation facilities, including the safe disposal of human waste and access to handwashing with soap and water at home (Global Sanitation, 2024).

5.3 TRYING FREE WATER POLICY IN SOUTH AFRICA

The South African constitution (Act 108 of 1996), Section 27(1)(b) states that 'Everyone has the right to have access to sufficient water'. In 2000, the Government introduced the ambitious Free Basic Water policy, which allows every household to get six with 6000 litres of potable water per month within 200 metres of the home at no cost (Farrar & Rivett, 2012).

According to Farrar and Rivett (2012), a national average of roughly 86% of households in the country were served with safe water by 2012. This was a significant improvement on the service delivery statistics from the apartheid era, where only 66% of urban households and 39% of rural households received this level of service.

Hall *et al.* (2006) noted that the Free Basic Water policy, however, reached more of those who were not poor than the poor, since the poor are less likely to have access to water services. Poorer and weaker municipalities are administratively and financially less able to implement this policy as effectively as better resourced municipalities. The South African case shows that implementing any type of free water policy is challenging and its implementation requires particularly strong capacity at all levels.

5.4 GOOD GOVERNANCE

A fundamental starting point in developing water services is good governance (see introduction to chapter three) and its related principles as shown in Figure 5.7. Governance needs to be open and accountable, coherent and integrative, responsive and sustainable, and equitable and ethical. Furthermore, good governance is to be consensus-oriented, participatory, and inclusive. Laws should be enforced and governance should be effective and efficient. Although Finland is among the countries with the highest transparency (Transparency International), there is still room for improvement. Bribery as such hardly

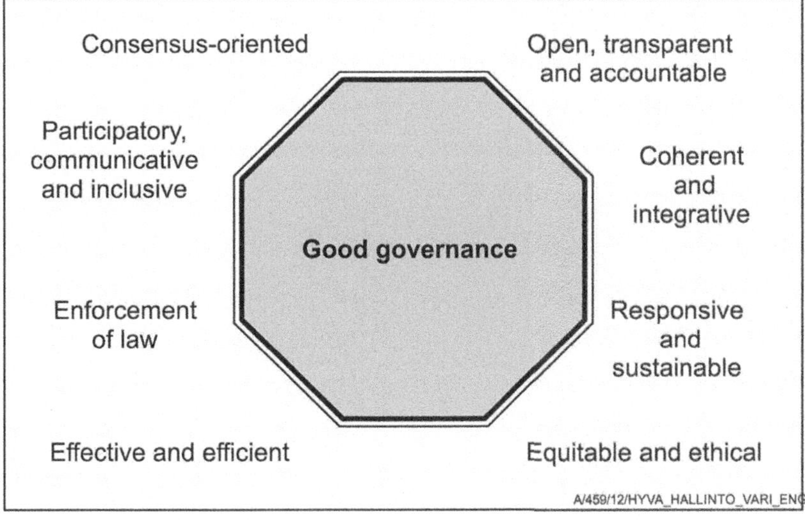

Figure 5.7 Dimensions of good governance.
Source: Seppälä (2004, p. 27), modified from: Rogers and Hall (2002); UNESCAP.

occurs in Finland, while it is more difficult to reveal less visible and non-transparent networks. (Salminen & Ikola-Norbacka, 2010) On the other hand, a small country has the advantage of an agile and lean administration although legal incompetence may sometimes be a constraint.

Several sources indicate that in low-income economies the constraints on water and sanitation services and inefficiency of sector organisations are essentially governance problems. Lack of good governance principles is actually one of the root causes of all major constraints within our societies.

According to Stockholm International Water Institute, water governance refers to the political, social, economic and administrative systems that influence water management and its use. Water governance is essentially about who gets what water, when and how, and who has the right to water, its related services and their benefits. (SIWI, n.d.)

OECD (2011b) pointed out that 'managing water for all is not only a question of resources availability and money, but equally a matter of good governance.' Furthermore, water is essentially a local issue involving many stakeholders at various levels. Major donors and international financial institutions are also increasingly requiring good governance as a condition for their support.

Myth 20: Once polluted, rivers and lakes are lost forever

Figure 5.8 While some argue that rivers, once polluted, are gone forever, others argue that a lot can be done to restore water bodies and control pollution.
Illustration: Pertti Väyrynen.

It is true that in some parts of the world too much water is used and is often polluted without proper conservation and pollution control as shown in the data relating to Myth 11. Some years ago, an award-winning documentary film presented the water quality problems existing in some major South-Asian rivers. Seen through the eyes of a sanitary engineer, the film gave the impression that river water would be lost for ever; the options for treating wastewaters before discharging them into the water body, or for renovating the waters was not mentioned (Figure 5.8).

Accordingly, during the Nordic summer holidays, the main TV channel always features a research vessel going to the Baltic Sea to take surface water samples. While this is naturally a step forward, we seldom hear news about what can be done in practice to implement water pollution control.

To show evidence of proper water pollution control, a short historical outline is presented here. During the last two centuries, intensive international efforts have been made to reach sufficient water quantity and quality. Simultaneously, human life expectancy has increased all over the world at unprecedented rates. This dramatic growth is undoubtedly due to improvements in water quality, but also other factors of improved water technologies and engineering. In addition, improved sanitation, personal hygiene, advances in medicine, and better living conditions have supported this development (Angelakis *et al.*, 2021, p. 1).

Water technology has evolved together with sanitation, personal hygiene, medicine, pesticides, education, food production and safety, housing, transportation, and communication technology. With a general rise in living standards, a combination of these elements has reduced morbidity and mortality. Yet the impact of these factors on life expectancy has also been counteracted by other circumstances such as overcrowding, unhealthy life habits, and ageing, particularly in Western countries (Angelakis *et al.*, 2021, p. 3).

Before the 1970s and 1980s, modern wastewater treatment and management were not generally developed and used. During more recent decades, water pollution control with modern wastewater treatment has become widely used in high-income countries. According to UN Habitat and WHO (2021), 56% of global household wastewater flows were safely treated, although the share of untreated wastewater is still far too high in many parts of the world (Myth 11).

In industrialised countries in the nineteenth and early twentieth centuries, positive long-term development of water quality and overall human health was observed. In 1900, the global average lifespan was just 31 years of age, and less than 50 years even in developed countries. By the mid-twentieth century, the average life expectancy had risen to 48 years. By the end of the twentieth century, the average lifespan reached 65 years, and in some countries this had risen to more than 80 years (Dattani *et al.*, 2023).

The history of the problem of wastewater in Tampere, Finland is briefly described below. Due to the inadequate service standard of the first low pressure system, the City Council in 1885 opted for a high pressure water system. However, to save costs, the proposed slow sand filters were excluded, even though the planner, Carl Hausen, is reported to have said: 'You will live to regret this' (Juuti, 2001, p. 126). This saving did indeed prove costly since an upstream sewage discharge point was located near the water intake. This resulted in typhus epidemics in 1908–09, killing 53 people, and an even more fatal one in 1915–16 with almost 300 fatalities (Figure 5.9).

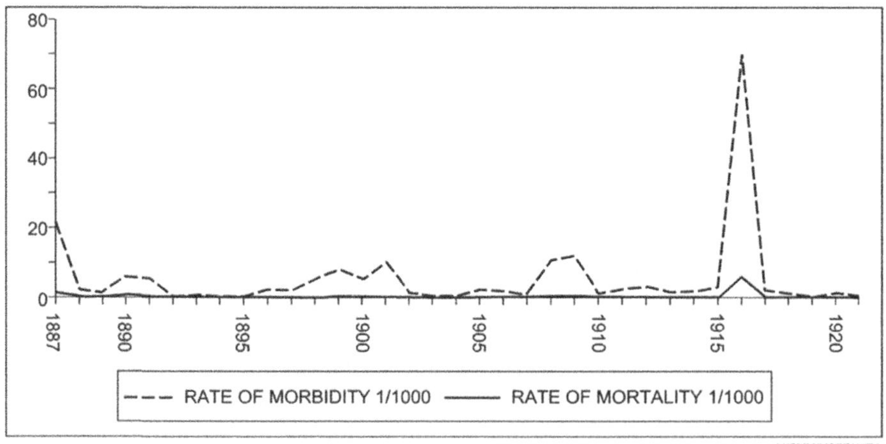

Figure 5.9 Typhoid epidemics and related morbidity and mortality rates in Tampere, 1887–1920.
Source: Koskinen (1998, p. 90), cited by Katko and Juuti (2007, p. 22).

It has been estimated that the immediate and direct hospital and health care costs due to cutting costs in this way were four times the potential savings made, in addition to the impact of the lives lost on citizens and the local economy (Juuti, 2001, pp. 245–246). In 1917, the water intake pipe was lengthened and chlorination began. These changes ended typhus epidemics.

> **REALITY**: Even heavily polluted waters, such as lakes and rivers, can be cleaned up over time if there is sufficient will. Searching to make savings at the expense of health and safety is dangerous in all situations and at all times.

However, since this era of water-related epidemics in Finland, it took over 60 years to introduce proper water pollution control and wastewater treatment in the city (Katko & Juuti, 2007, pp. 35–39). The pollution of the raw water source for Tampere, Lake Näsijärvi, upstream of the city, became increasingly serious during the subsequent years. Wastewater treatment was considered in the 1920s but this was found to be too expensive. In 1954, a general plan for sewerage systems was completed that divided the city into two sewerage areas using gravity where possible. In 1962, the first wastewater treatment plant with biologically activated sludge and a digestor for sludge treatment was completed for the west of Tampere.

Since 1972, the wastewaters of central and eastern Tampere have been diverted to a larger treatment plant close to the centre on another lake downstream of the city. Initially, this plant used mechanical treatment, but was upgraded with chemical precipitation in 1976 and biological-chemical treatment with simultaneous precipitation in 1982 (Katko & Juuti, 2007, pp. 35–37). The plant was partly delayed due to arguments over whether the sulphite pulp mill upstream should also treat their wastewater. In 1971, the pulp mill began to treat its wastewaters mechanically and in 1985, the mill was replaced by the country's first chemi-thermo-mechanical pulp mill. Biologically activated sludge treatment for wastewaters was also introduced, which decreased the wastewater load by 90% and increased external energy consumption dramatically. The mill was closed in 2008.

In addition to water pollution control and wastewater treatment, as elaborated in Myth 10, we can try to restore water bodies by various means. The 1992 European Commission (EC) Habitats Directive and the 2000 EC Water Framework Directive made river restoration a fundamental part of river management in Europe by requiring countries to improve the ecological status of their rivers (The River Restoration Centre). The removal of old dams no longer in active use or serving the purpose for which they were originally built, can be considered (American rivers, n.d.). Such interests are found in many parts of the world (e.g. Lima *et al.*, (2024) in Brazil; Dong (2024) in China; and Farguell *et al.* (2024) in Spain). For lakes, several renovation options are also available (Gann *et al.*, 2019).

In summary, in managing water bodies and controlling water pollution, a few principles are important relating to waterborne sanitation. First, a proper site for wastewater effluents should be selected, preferably downstream of any human settlements. Second, wastewaters should be treated as efficiently as possible, preferably starting from point sources of human settlements and industries and later from non-point sources. In addition, it is possible to try to renovate water bodies by various means.

Myth 21: Other countries' water problems don't concern us

Figure 5.10 We have only one world so let us keep it as clean and sustainable as possible. *Illustration:* Pertti Väyrynen.

Occasionally, one may encounter arguments that the problems of other nations or regions are of no concern to us; critics have presented claims, often lacking evidence, that development cooperation and international cooperation as a whole are ineffective. So what does this argument mean in terms of water services? The coronavirus in 2019 (COVID-19), defined as a pandemic from 11 March 2020 till 5 May 2023, is an example of joint international challenges that cannot be solved without international collaboration.

5.5 GLOBAL CHALLENGES OF WATER SERVICES

The introduction of clean water and sewage disposal – 'the sanitary revolution' – was chosen as the most important medical milestone to have occurred since 1840 by more than 11,300 readers of the British Medical Journal (Ferriman, 2007). This provides a useful historical record showing the importance of water services.

In 2016, the World Economic Forum assessed water related risks as one of the greatest global risks and threats in addition to climate change and weapons of mass destruction (WEF, 2016). The Global Risks Report of 2024 stated that over a two-year horizon, the most significant risks are AI-generated misinformation and disinformation, extreme weather events, societal polarisation, and cyber insecurity. Respectively, over this decade the biggest risks are: extreme weather events, critical changes to the Earth's systems, biodiversity loss and ecosystem collapse, and natural resource shortages (WEF, 2024). At least the first and the fourth issues are highly linked with water.

Water security goes far beyond whether we have too much or too little of this physical resource (Neo & Jha, 2023) (Figure 5.10). Water goes to the heart of every aspect of our development and wellbeing as people on a liveable planet. Enough water is needed of appropriate quality to keep us healthy, sustain our livelihoods, grow our economies, and protect our ecosystems. Water security covers a broad range of aspects from water-related disasters and waterborne diseases, conflicts over shared resources and governance challenges, to biodiversity and groundwater quality. Clearly, water is at the centre of this global crisis since nine out of 10 climate events are water related (Neo & Jha, 2023).

Selected water related challenges demonstrated by recent examples and studies are presented below. According to McGranahan *et al.* (2023), climate change is putting low-lying coastal zones at increased risk. They estimated that roughly 10% of the world's population and an even higher share of urban populations are located in coastal areas at below 10 metres in elevation, with Asia dominating these statistics.

Castro *et al.* (2023) explored the sanitary conditions of the third largest informal settlement in Brazil, including surface degradation, the lack of sanitation, or disasters related to climate events. In total, 35% of respondents said they release domestic sewage directly into the river near their homes, while 83% of them disposed urban solid waste inappropriately.

In Myth 19, several estimates are presented showing the health costs due to the lack of safe water, sanitation and health (WASH) and the potential benefits to be gained.

In developing economies, many water distribution systems are often unreliable. In Mexico City, many households store water in sealed rooftop tanks known as *tinacos*. Baisa *et al.* (2008) estimated that most of the potential distributional inefficiencies could be eliminated by making the frequency of deliveries the same across households rather than haphazard deliveries. This would require neither costly investments in infrastructure nor price increases (Baisa *et al.*, 2008).

The State of the World's Children 2024 presents three future scenarios: (i) business-as-usual trendlines; (ii) delayed development leading to greater inequality and environmental degradation; and (iii) accelerated development with a more inclusive and sustainable path (UNICEF, 2024). Scenario three should be in everyone's interests.

Hukka *et al.* (2018) pinpointed that in order to achieve the SDG targets 6.1 and 6.2 in non-OECD countries, in addition to making the required rehabilitation and replacement investments in OECD countries, a major increase – as much as threefold – in funding should be considered.

Climate change is visible through disturbances in the hydrological cycle when extreme events increase through heavier rainfalls and longer droughts. As Taalas (2023) pointed out: 'Climate change is the defining challenge of our time. How we respond to that challenge will determine the future of our planet and of our children and grandchildren'.

Many water management issues are wicked in nature as pointed out by Freeman (2000). Managing wicked problems requires changing the questions, managing uncertainty, and creating resilience. While this does not solve existing problems, it does drive progress to a desired future state (Beutler, 2016).

More collaborative governance and approaches for natural resources management and urban planning have been called for (e.g. by Koebele *et al.*, 2024; Kurki, 2016). Such approaches apply to water services, particularly to large scale projects. Through co-creation it is possible to practise 'slow thinking' – to focus on something long enough to have more positive and constructive views than would be the case in fast thinking as elaborated by Kahneman (2011).

The COVID-19 pandemic reminded us about the fundamental responsibility of the public sector in society. If industrial and logistical activities are in competition through overly complicated arrangements, security of supply will suffer. In problematic situations responsibilities are not clear. The COVID-19 pandemic was not the first, nor will it be the last. For this reason, improved water services are needed everywhere which will benefit all.

REALITY: International phenomena, such as climate refugees, inevitably affect any country. Developed economies need to share their experiences and positively impact global practices through development cooperation and exporting activities. One of the best ways is to invest in the education of experts in developing economies.

As a wider aim, it is justifiable to improve living conditions through sustainable development (health, education, water and sanitation) as this has been found to reduce the risk of becoming radicalised. One of the first ways of doing this are to honour citizen's rights of access to water services – clean water and adequate sanitation. To prevent pandemics, it is very important to be able to wash hands with hot water.

Interdependencies between and threats to various societal functions are illustrated in Figure 5.11, which originates from before the COVID-19 outbreak. It highlights the challenges of criminality and terrorism, global logistical disturbances, production chain threats to information systems and cyber attacks, as well as extreme weather conditions and energy shortages. Within this framework, Pekki (2019) identified a hierarchy of interdependent societal functions, starting with energy, followed by water and wastewater services (Figure 5.11).

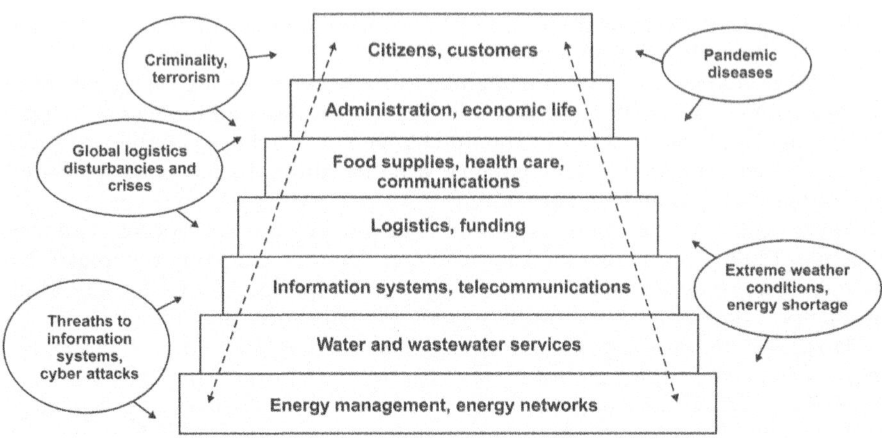

Figure 5.11 Interdependencies and threats of societal functions. *Source:* Pekki (2019, modified by the authors).

An example of this is the new Food and Water 2030 programme launched in December 2024 (NESA, 2024). Food and water supply form strategic focus areas since they have major impact(s) on the functioning of society, especially civil society, and in crisis situations.

5.6 TOWARDS BETTER COOPERATION

In the late 1960s, some international organisations, including the World Health Organization (WHO), the United Nations Children's Fund (UNICEF), and the International Development Research Centre (IDRC) in Canada, were already supporting the provision of rural drinking water supplies in low-income countries. In 1977, the World Water Conference in Mar del Plata, Argentina, adopted – based on the recommendations of the 1976 United Nations Habitat Conference – a declaration that initiated a new era in international cooperation for improved drinking water supplies and sanitation in the developing world. The International Drinking Water Supply and Sanitation Decade (IDWSSD) 1981–90 emerged from this declaration (Black, 1998).

Since then, successive international declarations and eras have been devoted, formulated, agreed and implemented for improving global access to safe water and sanitation – on the basis of international cooperation. In 2015, all 193 Member States of the United Nations General Assembly unanimously agreed their commitment to *Transforming our world: The 2030 Agenda for Sustainable Development* (the 2030 Agenda). The 2030 Agenda is described as a plan of action for people, planet and prosperity. At its heart are the 17 SDGs and 169 global targets, which are an urgent call for action by all countries in a global partnership (UNDESA, 2025).

The establishment of SDG6: Ensure availability and sustainable management of water and sanitation for all reflects the increased attention on water and sanitation issues in the global political agenda. Both target 6.a: By 2030, expand international cooperation and capacity-building support to developing countries in water- and sanitation-related activities and programmes, including water harvesting, desalination, water efficiency, wastewater treatment, recycling and reuse technologies and target 17.9: Enhance international support for implementing effective and targeted capacity-building in developing countries to support national plans to implement all the sustainable development goals, including through North-South, South-South and triangular cooperation call for increased international cooperation, and in particular, for capacity-building.

This is based on the findings that adequate human resource development for the water and sanitation sector has long been recognised as a priority issue (Cavill & Saywell, 2009). Furthermore, developed countries have 20 to 50 scientists and engineers per 10 000 population, compared to about five scientists and engineers on average in developing countries, and to one or less scientist or engineer for some poorer African countries (Marjoram, 2010). The shortage of engineers in developing countries is hampering development, especially for the poorest citizens who are still lacking adequate water and sanitation services. Rees *et al.* (2008) also stressed that water management reform and attempts to improve the financial viability of water service producers in the sector will not succeed unless considerable efforts are also made to increase human capacity.

An UN-Water Global Analysis and Assessment of Sanitation and Drinking-Water report showed that in many countries, policies and programmes have far too little emphasis on ensuring adequate financial and human resources to both sustain the existing infrastructure and expand access to sanitation, drinking-water and hygiene services. Many of the governments, reporting inadequate funding allocations for WASH (water, sanitation and hygiene), also point to a poor absorption capacity – that is, difficulties in spending the limited funds that are received (WHO, 2012b).

A recent World Bank's assessment of global public spending shows that one of the main challenges of the water sector is its low budget execution rate, compared to the human development sector and nearly every other infrastructure sector. These rates averaged at about 72% during 2009–2020, indicating that 28% of allocated funds go unspent. Low budget execution rates point to systemic constraints on the sector's absorptive capacity, which in turn is due to a range of institutional, governance, project management, and political economy factors. In many countries, the mandate to provide water and sanitation services lies with the subnational governments, which rely heavily on budgetary transfers from federal or provincial governments. Yet at the same time, they typically have limited capacity to prepare project plans and budget proposals or to implement projects on an annual- or medium-term basis. This limitation affects their ability to receive resources from higher-level governments (Joseph *et al.*, 2024).

According to Makengo *et al.* (2021), access to water may well be a major stake for conflicts in this 21st century. The need for water diplomacy between independent nations is obvious and necessary. Yet regarding water services, the focus of collaboration and co-creation should be largely at local levels, especially in municipalities or respective bodies where provision and production takes place. This part of water diplomacy has so far been largely ignored.

Considering the foreseen challenges and needs, it would be well justified to increase international development collaboration and capacity building by supporting universities especially in low-income countries through sharing knowledge, education and research cooperation. Improved water services – including sanitation – makes it possible to mitigate wicked problems and adapt to changing conditions.

The former director of WMO, Taalas (2024) considered that in spite of its shortcomings, the UN aims to operate for the benefit of mankind through its projects and words. The agency acts also as a guardian of morality and tries to control governments, which do not promote the best interests of their citizens.

In a wider context, we can refer to Nobel Peace Laureate, Martti Ahtisaari, who noted that 'the severe challenge for our societies in the 21st century is not only how we manage to create wealth but also how we use it' (Merikallio & Ruokanen, 2015). As the UN Secretary-General, António Guterres, stated: 'In turbulent times, work of the United Nations is more necessary than ever' (UN News, 2024).

5.7 DISCUSSION QUESTIONS

(1) How could thinking about and discussing long-term thinking and development be promoted?
(2) Why are the poorest people often obliged to pay the highest price relatively for water – especially drinking water?
(3) How do you see the contradictions between nature conservation and active environmental protection in terms of, for example, water pollution control?
(4) What kinds of approaches could be most feasible to promote and support water services in the Southern hemisphere?

REFERENCES

Acemoglu D. and Johnson S. (2023). Power and Progress: Our Thousand Year Struggle Over Technology and Prosperity. John Murray Press, London, UK.

Acemoglu D. and Robinson J.-A. (2019). The Narrow Corridor: States, Societies, and the Fate of Liberty. Penguin Publishers, New York.

American rivers (n.d.). Life depends on rivers. https://www.americanrivers.org/ (accessed 22 March 2025).

Angelakis A. N., Vuorinen H. S., Nikolaidis C., Juuti P. S., Katko T. S., Juuti R. P., Zhang J. and Samonis G. (2021). Water quality and life expectancy: parallel courses in time. *Water*, **13**, 752, https://doi.org/10.3390/w13060752

Antila K., Katko T. S. and Mattila H. (2013). Technology development theories and water services evolution. In: Water Services Management and Governance: Past Lessons for a Sustainable Future, T. Katko, P. Juuti and K. Schwartz (eds), IWA Publishing, London, UK, pp. 13–26. https://library.oapen.org/handle/20.500.12657/25810 (accessed 22 March 2025).

Baisa, B., Davis L. W., Salant S. W. and Wilcox W. (2008). The welfare costs of unreliable water service. *Journal of Development Economics*, **92**(1), 1–12, https://doi.org/10.1016/j.jdeveco.2008.09.010

Beutler L. (2016). What to Do about Wicked Water Problems. University of Arizona. https://wrrc.arizona.edu/wicked-water-problems (accessed 22 March 2025).

Bezahler A. (2024). Dams in intrastate ethnic conflict: a comparison of Turkey and Ethiopia. *World Water Policy*, **10**, 1044–1065, https://doi.org/10.1002/wwp2.12233

Black M. (1998). Learning What Works. A 20 Year Retrospective View on International Water and Sanitation Cooperation: 1978–1988, pp. 4, 6. https://www.ircwash.org/resources/learning-what-works-20-year-retrospective-view-international-water-and-sanitation (accessed 22 March 2025).

Burek P., Satoh Y., Fischer G., Kahil M. T., Scherzer A., Tramberend S., Nava L. F., Wada Y., Eisner S., Flörke M., Hanasaki N., Magnuszewski P., Cosgrove B. and Wiberg D. (2016). Water Futures and Solution: Fast Track Initiative (Final Report), IIASA Working Paper. International Institute for Applied Systems Analysis (IIASA), Laxenburg, Austria. https://pure.iiasa.ac.at/id/eprint/13008/ (accessed 22 March 2025).

Castro F. A. B., de Sá Salomão A. L. and Netto A. T. (2023). Sanitary conditions of the third largest informal settlement in Brazil. *Journal of Water, Sanitation and Hygiene for Development*, **13**(12), 962, https://doi.org/10.2166/washdev.2023.152

Cavill S. and Saywell D. (2009). The capacity gap in the water and sanitation sector. The 34th WEDC International Conference, Addis Ababa, Ethiopia, p. 1. http://wedc.lboro.ac.uk/resources/conference/34/Cavill_S_-_335.pdf (accessed 15 Dec 2024).

Critical 5 (2014). Forging a Common Understanding for Critical Infrastructure: Shared Narrative. https://www.cisa.gov/sites/default/files/publications/critical-five-shared-narrative-critical-infrastructure-2014-508.pdf (accessed 22 March 2025).

Dattani S., Rodés-Guirao L., Ritchie H., Ortiz-Ospina E. and Roser M. (2023). Life Expectancy. Published online at OurWorldinData.org. https://ourworldindata.org/life-expectancy (accessed 22 March 2025).

Dong Y. (2024). Revealing the Willingness-to-pay for River Restoration in China: a Meta-analysis. *Environment, Development and Sustainability*, https://doi.org/10.1007/s10668-024-05787-9

European Association of Public Water Operators (2019). The Public Water Services of The Future - Within Society for Sustainability, p. 7. https://www.aquapublica.eu/document/public-water-services-future (accessed 22 March 2025).

Farguell J., Chavez J. and Ochoa L. (2024). Assessment of a process-based urban river restoration using biological and hydro-geomorphological indicators. The Congost River at Granollers (Catalonia, Spain). *Journal of Environmental Management*, **369**, 122424, https://doi.org/10.1016/j.jenvman.2024.122424

Farrar L. and Rivett U. (2012). Is South Africa's free basic water policy working? A quantitative analysis of the effectiveness of the policy implementation. Department of Civil Engineering, University of Cape Town. WISA 2012 – P030, p. 1. https://wisa.org.za/wp-content/uploads/2018/12/WISA2012-P030.pdf (accessed 22 March 2025).

Ferriman A. (2007). BMJ readers choose the 'sanitary revolution' as greatest medical advance since 1840. *BMJ*, **334**, 111, https://pmc.ncbi.nlm.nih.gov/articles/PMC1779856/ (accessed 22 March 2025).

Freeman D. M. (2000). Wicked water problems: Sociology and local water organizations in addressing water resources policy. *Journal of the American Water Resources Association*, **36**(3), 483–491.

Gann G. D., McDonald T., Walder B., Aronson J., Nelson C. R, Jonson J, Hallett J. G., Eisenberg C, Guariguata M. R., Liu J., Hua F., Echeverría C., Gonzales E., Shaw N., Decleer K., and Dixon K. W. (2019). International principles and standards for the practice of ecological restoration. Second edition. *Restoration Ecology*, **27**(S1), https://doi.org/10.1111/rec.13035

Gleick P. H. (2002). Dirty Water: Estimated Deaths from Water-Related Diseases 2000–2020, p. 4. https://pacinst.org/wp-content/uploads/2013/02/water_related_deaths_report3.pdf (accessed 22 March 2025).

Gleick P. (2023). The Three Ages of Water: Prehistoric Past, Imperiled Present, and a Hope for the Future. Public Affairs, New York, USA.

Gleick P. H. (2024). Moving to a new age of water. *Environment: Science and Policy for Sustainable Development*, **66**(1), 42–45, https://doi.org/10.1080/00139157.2023.2269046

Global Sanitation (2024). Global Water, Sanitation and Hygiene (WASH). 23 Oct 2024. https://www.cdc.gov/global-water-sanitation-hygiene/about/about-global-sanitation.html (accessed 22 March 2025).

Hall K., Leatt A. and Monson J. (2006). Accommodating the poor: The Free Basic Water policy and the Housing Subsidy Scheme. South African Child Gauge. Part Two: Children and Poverty. pp. 57–62. https://ci.uct.ac.za/sites/default/files/content_migration/health_uct_ac_za/533/files/gauge2006_accom.pdf (accessed 22 March 2025).

Heymans C., Rolfe E., David E. and Riley S. (2016). Providing Water to Poor People in African Cities Effectively: Lessons from Utility Reforms. World Bank, Washington, DC. https://openknowledge.worldbank.org/entities/publication/d115f93d-0bda-5210-b33b-724ddf030f71 (accessed 22 March 2025).

Hukka J. J. and Katko T. S. (2021). Towards Sustainable Water Services: Subsidiarity, Multi-level Governance and Resilience for Building Viable Water Utilities. CADWES Publications. http://www.cadwes.com/publications/others/ (accessed 22 March 2025).

Hukka J. J., Mattila H. E. and Katko T. S. (2018). The Quest for SDG6 and Community Water Services Resilience – Factor 3? IWA World Congress & Exhibition, 16–21 Sept 2018, Tokyo, Japan.

Hutton G. (2018). Global benefits and costs of achieving universal coverage of basic water and sanitation services as part of the 2030 Agenda for Sustainable Development. In: Prioritizing Development: A Cost Benefit Analysis of the United Nations' Sustainable Development Goals, B. Lomborg (ed.), Cambridge University Press, Cambridge, pp. 422–445, pp. 16–20. https://assets.cambridge.org/97811084/15453/frontmatter/9781108415453_frontmatter.pdf (accessed 22 March 2025).

Hutton G. and Varughese M. (2016). The Costs of Meeting the 2030 Sustainable Development Goal Targets on Drinking Water, Sanitation, and Hygiene. Water and Sanitation Program: Technical Paper 103171, International Bank for Reconstruction and Development/World Bank and Water and Sanitation Program (WSP), p. xi. https://www.worldbank.org/en/topic/water/publication/the-costs-of-meeting-the-2030-sustainable-development-goal-targets-on-drinking-water-sanitation-and-hygiene (accessed 22 March 2025).

International Law Association (2004). Water Resources Law. The Berlin Rules on Water Resources. Berlin Conference 2004, p. 22. https://www.internationalwaterlaw.org/documents/intldocs/ILA/ILA_Berlin_Rules-2004.pdf (accessed 22 March 2025).

Joseph G., Hoo Y. R., Wang Q., Bahuguna A. and Andres L. (2024). Funding a Water-Secure Future: An Assessment of Global Public Spending. World Bank, Washington, DC., p. xxvii, 250–251.

Juuti P. (2001). Kaupunki ja vesi. Tampereen vesihuollon ympäristöhistoria 1835–1921. Summary: City and Water. Doctoral dissertation, Dept. of History, University of Tampere. https://urn.fi/urn:isbn:951-44-5232-1 (accessed 22 March 2025).

Juuti P. S., Katko T. S. and Vuorinen H. S. (eds) 2007. Environmental History of Water – Global views on community water supply and sanitation. IWA Publishing, London. http://www.iwapublishing.com/template.cfm?name=isbn1843391104 (accessed 15 December 2024).

Kahneman D. (2011). Thinking, Fast and Slow. Paperback. Farrar, Strauss and Giroux, New York, pp. 282–283.

Kaivo-oja J. Y., Katko T. S. and Seppälä O. T. (2004). Seeking for convergence between history and futures research. *Futures*, **36**, 527–547, https://doi.org/10.1016/j.futures.2003.10.017

Kashiwase H. (2019). Open defecation nearly halved since 2000 but is still practiced by 670 million. 19 Nov 2019. https://blogs.worldbank.org/en/opendata/open-defecation-nearly-halved-2000-still-practiced-670-million (accessed 22 March 2025).

Kashiwase H. (2023). World Toilet Day: 420 million people are defecating outdoors. 17 Nov 2023. https://blogs.worldbank.org/en/opendata/world-toilet-day-420-million-people-are-defecating-outdoors (accessed 22 March 2025).

Katko T. S. (2016). Finnish Water Services – Experiences in Global Perspective. Finnish Water Utilities Association, Helsinki, Finland. Co-published E-book, IWA Publishing, London, 2017. www.finnishwaterservices.fi (accessed 15 December 2024).

Katko T. S. and Hukka J. J. (2021). Crisis and Water Services: How a 2007 Public Health Emergency in Finland Helped Shape Its Response to COVID-19. *Public Works Management & Policy*, **26**(1), 63–70.

Katko T. and Juuti S. (2007). Watering the city of Tampere from the mid-1800s to the 21st century. Tampere Water and International Water History Association. https://tampub.uta.fi/handle/10024/65709 (accessed 15 December 2024).

Katko T. S., Juuti P. S. and Hukka J. J. (2002). An early attempt to privatize – any lessons learnt? Research and technical note. *Water International*, **27**(2), 294–297, https://doi.org/10.1080/02508060208687

Katko T., Juuti P. and Rajala R. (2009). Writing the history of water services. *Physics and Chemistry of the Earth*, **34**, 156–163, p. 2, https://doi.org/10.1016/j.pce.2008.06.033

Katko T. S., Juuti P. S., Juuti R. P. and Väyrynen P. O. (2022). Vesihuollon myytit (The myths of water services). Vastapaino, Tampere, Finland, p. 166.

Koebele E. A., Méndez-Barrientos L. E., Nadeau N., Gerlak A. K. (2024). Beyond engagement: enhancing equity in collaborative water governance. *Wires Water*, **11**(2), e1687, https://doi.org/10.1002/wat2.1687

Koskinen M. (1998). Kun kulkutauti saapui Tampereelle. In: Ernomane vesitehdas – Tampereen vesilaitos 1835–1998, P. Juuti and T. Katko (eds), Tampereen kaupungin vesilaitos, Tampere, Finland, pp. 88–99, p. 90. https://trepo.tuni.fi/handle/10024/66324 (accessed 22 March 2025).

Kurki V. (2016). Negotiating groundwater governance: lessons from contentious aquifer recharge projects. Doctoral dissertation no. 1387, Tampere University, Finland, pp. ii–iii. https://trepo.tuni.fi//handle/10024/115229 (accessed 22 March 2025).

Lima A. F., Ferreira G. F., Veról A. P. and Miguez M. G. (2024). Prioritizing urban river restoration management practices: a cross-evaluation using the Criticality Index for Watershed Restoration (CIWR) and opportunity layers. *Land*, **13**, 2244, https://doi.org/10.3390/land13122244

Lombana Cordoba C., Saltiel G. and Perez Penalosa F. (2022). Utility of the Future, Taking Water and Sanitation Utilities Beyond the Next Level 2.0. A Methodology to Ignite Transformation in Water and Sanitation Utilities. World Bank, Washington, DC. https://documents.worldbank.org/en/publication/documents-reports/documentdetail/099325005112246075/p1655860d146090dc097dc0165740604044 (accessed 22 March 2025).

Makengo B. M., Molanga J. M., Mbutamuntu J. M., Londo P. K. and Nsiy T. M. K. (2021). Water: a major stake of conflicts in the twenty-first century. *Open Journal of Social Sciences*, **9**, 125–148, p. 125, https://doi.org/10.4236/jss.2021.911011

Marjoram T. (ed.) (2010). Engineering: issues, challenges and opportunities for development. UNESCO report, UNESCO, Paris, France, p. 311. https://www.acofi. edu.co/wp-content/uploads/2013/08/Issues-challenges.pdf (accessed 22 March 2025).

McGranahan G., Balk D., Colenbrander S., Engin H. and MaCManus K. (2023). Is rapid urbanization of low-elevation deltas undermining adaptation to climate change? A global review. *Environment & Urbanization*, **35**(2), 527–559, p. 527, https://doi. org/10.1177/09562478231192176

Merikallio K. and Ruokanen T. (2015). Ch. 6. Development Aid and the Foreign Ministry Fledging. In: The Mediator: A Biography of Martti Ahtisaari, K. Merikallio and T. Ruokanen (eds), 1st edn, Hurst & Company, London, UK, pp. 51–68.

Mumssen Y., Saltiel G. and Kingdom B. (2018). Aligning Institutions and Incentives for Sustainable Water Supply and Sanitation Services: Report of the Water Supply and Sanitation Global Solutions Group, Water Global Practice, World Bank, Washington, DC. https://documents.worldbank.org/en/publication/documents-reports/documentdetail/271871525756383450/aligning-institutions-and-incentives-for-sustainable-water-supply-and-sanitation-services (accessed 22 March 2025).

Naughton J. (2023). Power and Progress review – why the tech-equals-progress narrative must be challenged. Observer, 7 May 2023. https://www.theguardian.com/ books/2023/may/07/power-and-progress-daron-acemoglu-simon-johnson-review-formidable-demolition-of-the-technology-equals-progress-myth (accessed 22 March 2025).

Neo G. H. and Jha A. K. (2023). Why water security is our most urgent challenge today. World Economic Forum, 12 Oct 2023. https://www.weforum.org/stories/2023/10/ why-water-security-is-our-most-urgent-challenge-today/ (accessed 22 March 2025).

NESA (2024). National Emergency Supply Agency launches new Food and Water 2030 programme. 20 Dec 2024. https://www.huoltovarmuuskeskus.fi/en/a/ national-emergency-supply-agency-launches-new-food-and-water-2030-programme (accessed 22 March 2025).

North D. C. (2003). Understanding the Process of Economic Change. Forum Series on the Role of Institutions in Promoting Economic Growth. Mercatus Center. George Mason University, p. 12. https://pdf.usaid.gov/pdf_docs/Pnacx402.pdf (accessed 15 December 2024).

North Carolina State Water Infrastructure Authority (2017). North Carolina's Statewide Water and Wastewater Infrastructure Master Plan: The Road to Viability, p. 16. https://www.deq.nc.gov/water-infrastructure/statewide-water-and-wastewater-infrastructure-master-plan-2017/open (accessed 22 March 2025).

OECD (2011a). Benefits of Investing in Water and Sanitation: An OECD Perspective. OECD Publishing, Paris, p. 16. https://www.oecd.org/en/publications/benefits-of-investing-in-water-and-sanitation_9789264100817-en.html (accessed 22 March 2025).

OECD (2011b). Water Governance in OECD Countries: A Multi-level Approach, p. 3. www.oecd.org/gov/regional-policy/watergovernanceinoecdcountriesamulti-levelapproach.htm (accessed 22 March 2025).

OECD (2019). Good Governance for Critical Infrastructure Resilience. OECD Reviews of Risk Management Policies. OECD Publishing, Paris, France. https:// www.oecd.org/en/publications/good-governance-for-critical-infrastructure-resilience_02f0e5a0-en.html (accessed 22 March 2025).

Office of the Auditor-General, New Zealand (2014). Water and roads: Funding and management challenges, p. 16. https://oag.parliament.nz/2014/assets/docs/water-and-roads.pdf (accessed 22 March 2025).

Office of the Director of National Intelligence (2021). Structural Drivers of the Future. Environmental and Resource Trends: Water Insecurity Threatening Global Economic Growth, Political Stability. https://www.dni.gov/files/images/globalTrends/GT2040/NIC_2021-02489_Future_of_Water_18nov21_UNSOURCED.pdf (accessed 22 March 2025).

OHCHR (Office of the United Nations High Commissioner for Human Rights) (2010). The Right to Water. Fact Sheet No. 35. Geneva, Switzerland, p. 13. https://www.refworld.org/reference/themreport/ohchr/2010/en/76003 (accessed 22 March 2025).

Pekki J. (2019). Varautuminen strategisena ajatteluna – miten kriittinen infrastruktuuri kytkeytyy huoltovarmuuteen? (Preparedness as strategic thinking – how is critical infrastructure linked to security of supply), Tampere University, 23 Jan 2019.

Rees J. A., Winpenny J. and Hall A. W. (2008). Water Financing and Governance. Global Water Partnership Technical Committee (TEC) Background Papers No. 12, p. 44. http://www.gwp.org/Global/ToolBox/Publications/Background%20papers/12%20Water%20Financing%20and%20Governance%20(2008)%20English.pdf (accessed 22 March 2025).

Rogers P. and Hall A. W. (2002). Effective water governance, TEC Background paper no. 7, GWP. https://www.gwp.org/globalassets/global/toolbox/publications/background-papers/07-effective-water-governance-2003-english.pdf (accessed 22 March 2025).

Ruonavaara H. (2006). Historian polut ja teorian kartta eli miten tutkia tapahtumaketjuja sosiologisesti (Historical paths and road map: how to study chains of events in sociology). In: Historiallinen käänne. Johdatus pitkän aikavälin historian tutkimukseen, J. S. Saari (ed.), Gaudeamus, Helsinki, pp. 34–63.

Sadoff C. W., Hall, J. W., Grey D., Aerts J. C. J. H., Ait-Kadi M., Brown C., Cox A., Dadson S., Garrick D., Kelman J., McCornick P., Ringler C., Rosegrant M., Whittington D. and Wiberg D. (2015). Securing Water, Sustaining Growth. Report of the GWP/OECD Task Force on Water Security and Sustainable Growth, University of Oxford, UK, p. 22. https://www.gwp.org/globalassets/global/about-gwp/publications/the-global-dialogue/securing-water-sustaining-growth.pdf (accessed 22 March 2025).

Salminen A. and Ikola-Norbacka R. (2010). Trust, good governance and unethical actions in Finnish public administration. *IJPSM*, **23**(7), 647–668.

Sandelin S. (2017). Knowledge Management and Retention: A Case of a Water Utility in Finland. Doctoral dissertation no. 1476, Tampere University of Technology, Finland, pp. i–ii. https://urn.fi/URN:ISBN:978-952-15-3959-6 (accessed 22 March 2025).

Santayana G. (1905). The Life of Reason. From the series Great Ideas of Western Man. https://www.si.edu/object/those-who-cannot-remember-past-are-condemned-repeat-it-george-santayana-life-reason-1905-series%3Asaam_1984.124.194 (accessed 22 March 2025).

Sedlak D. (2014). Water 4.0. The Past, Present, and Future of the World's Most Vital Resource. Yale Univ. Press, New Haven, CT, p. 238.

Sedlak D. (2021). The need for a 'Next Path' on water. The Source, 23 July 2021. https://www.thesourcemagazine.org/the-need-for-a-next-path-on-water/ (accessed 22 March 2025).

Sedlak D. (2024). Water for All: Global solutions for a changing climate. Maven's Notebook. 12 Sept 2024. https://mavensnotebook.com/2024/09/12/david-sedlak-water-for-all-global-solutions-for-a-changing-climate/ (accessed 22 March 2025).

Seppälä O. T. (2004). Visionary management in water services: Reform and development of institutional frameworks. Doctoral dissertation no. 457, Tampere University of Technology, Finland.

SIWI (n.d.). Water governance. https://siwi.org/why-water/water-governance/ (accessed 22 March 2025).

Skotnes R. (2016). Division of Cyber Safety and Security Responsibilities Between Control System Owners and Suppliers. In: Critical Infrastructure Protection, X. M. Rice and S. Shenoi (eds), IFIP International Federation for Information Processing. AICT Advances in Information and Communication Technology 485. Springer, Switzerland, p. 133.

Stiglitz J. E. (2023). Inequality and Democracy. Project Syndicate, 31 Aug 2023. https://www.project-syndicate.org/commentary/inequality-source-of-lost-confidence-in-liberal-democracy-by-joseph-e-stiglitz-2023-08 (accessed 22 March 2025).

Taalas P. (2023). The Future of Weather, Climate and Water across Generations. *UN Chronicle*, 23 March 2023. https://www.un.org/en/un-chronicle/future-weather-climate-and-water-across-generations (accessed 22 March 2025).

Taalas P. (2024) Maailmanparantajat. Miltä YK näyttää sisältäpäin? (Healer of the world. What does UN look like from inside) Tammi, Helsinki, Finland, pp. 150–153.

The River Restoration Centre. What is river restoration? Esmée Fairbairn Foundation. https://www.therrc.co.uk/assets/general/Training/esmee/what_is_river_restoration_final.pdf (accessed 22 March 2025).

Too E. G. (2012). Strategic Infrastructure Asset Management: The Way Forward. In: Engineering Asset Management and Infrastructure Sustainability, J. Mathew, L. Ma, A. Tan, M. Weijnen and J. Lee (eds), Springer, London, p. 2, https://doi.org/10.1007/978-0-85729-493-7_73

Transparency International. Corruption Perceptions Index. https://www.transparency.org/en/cpi/2023 (accessed 22 March 2025).

UN (2021). The United Nations World Water Development Report 2021: Valuing Water, UNESCO, Paris, p. 18. https://www.unesco.org/reports/wwdr/2021/en (accessed 15 Dec 2024).

UNDESA (United Nations Department of Economic and Social Affairs) (2025). Sustainable Development. The 17 Goals. https://sdgs.un.org/goals (accessed 22 March 2025).

UNEP (2012). Measuring water use in a green economy. A Report of the Working Group on Water Efficiency to the International Resource Panel, J. McGlade, B. Werner, M. Young, M. Matlock, D. Jefferies, G. Sonnemann, M. Aldaya, S. Pfister, M. Berger, C. Farell, K. Hyde, M. Wackernagel, A. Hoekstra, R. Mathews, J. Liu, E. Ercin, J. L. Weber, A. Alfieri, R. Martinez-Lagunes, B. Edens, P. Schulte, S. von Wirén-Lehr and D. Gee, pp. 14–15. https://www.resourcepanel.org/reports/measuring-water-use-green-economy (accessed 22 March 2025).

UNESCAP. What is good governance? https://www.unescap.org/sites/default/files/good-governance.pdf (accessed 22 March 2025).

UNESCO (2023). The United Nations World Water Development Report 2023: partnerships and cooperation for water; executive summary, UNESCO, Paris, France, p. 3. https://unesdoc.unesco.org/ark:/48223/pf0000384657 (accessed 22 March 2025).

UN Habitat and WHO (2021). Progress on wastewater treatment – Global status and acceleration needs for SDG indicator 6.3.1. Geneva, Switzerland. https://unhabitat.org/progress-on-wastewater-treatment-%E2%80%93-2021-update (accessed 22 March 2025).

UNICEF (2024). The State of the World's Children 2024. The future of childhood in a changing world. Three future scenarios. https://www.unicef.org/reports/

state-of-worlds-children/2024?utm_campaign=SOWC%20statistical%20 compendium&utm_medium=email&utm_source=Mailjet#futures (accessed 22 March 2025).
UNICEF and WHO (2023). Progress on household drinking water, sanitation and hygiene 2000–2022: special focus on gender. https://www.who.int/publications/m/ item/progress-on-household-drinking-water--sanitation-and-hygiene-2000-2022--- special-focus-on-gender (accessed 22 March 2025).
United Nations and the World Bank (2018). Making every drop count. An Agenda for Water Action. High Level Panel on Water Outcome Document, p. 26. https:// reliefweb.int/report/world/making-every-drop-count-agenda-water-action-high- level-panel-water-outcome-document-14 (accessed 22 March 2025).
UN News (2024). In turbulent times, 'work of the United Nations is more necessary than ever': Guterres. https://www.un.org/en/desa/turbulent-times-%E2%80%98work- united-nations-more-necessary-ever%E2%80%99-guterres (accessed 22 March 2025).
UN-Water (2015). A compilation of aspects on the means of implementation: water and sanitation. A look at Goal 6 and Goal 17. Advanced draft copy – 24 April 2015. pp.14–15. https://www.unwater.org/publications/means-implementation-focus- sustainable-development-goals-6-and-17 (accessed 22 March 2025).
UN-Water (2018). SDG 6 Synthesis Report 2018 on Water and Sanitation, United Nations, New York, p. 26. https://www.unwater.org/sites/default/files/app/ uploads/2018/12/SDG6_SynthesisReport2018_WaterandSanitation_04122018. pdf (accessed 22 March 2025).
UNW-DPAC and WSSCC (2011). The Human Right to Water and Sanitation. Media Brief. https://sswm.info/node/2653 (accessed 22 March 2025).
WEF (World Economic Forum) (2016). The Global Risks Report 2016. Geneva, Switzerland, Figure 1. https://www.weforum.org/reports/the-global-risks- report-2016 (accessed 22 March 2025).
WEF (World Economic Forum) (2024). The Global Risks Report. Summary, Cologny/ Geneva, Switzerland, p. 8. https://www3.weforum.org/docs/WEF_The_Global_ Risks_Report_2024.pdf (accessed 22 March 2025).
Whittington D. (2015). Benefits and Costs of the Water Sanitation and Hygiene Targets for the Post-2015 Development Agenda. Post-2015 Consensus. Water and Sanitation Perspective Paper. Copenhagen Consensus Center, Denmark, p. 10. https:// copenhagenconsensus.com/sites/default/files/was_perspective_-_whittington_0. pdf (accessed 22 March 2025).
WHO (2012a). Global Costs and Benefits of Drinking-Water Supply and Sanitation Interventions to Reach the MDG Target and Universal Coverage. Geneva, Switzerland, p. 47. https://www.who.int/publications/i/item/WHO-HSE-WSH-12.01 (accessed 22 March 2025).
WHO (2012b). GLAAS 2012 Report. UN-Water Global Analysis and Assessment of Sanitation and Drinking-Water: The Challenge of Extending and Sustaining Services, p. 4. http://www.un.org/waterforlifedecade/pdf/glaas_report_2012_eng. pdf (accessed 15 Dec 2024).
World Bank (2016). High and Dry: Climate Change, Water, and the Economy. World Bank, Washington, DC. https://www.worldbank.org/en/topic/water/publication/ high-and-dry-climate-change-water-and-the-economy (accessed 22 March 2025).
WWAP (2012). The United Nations World Water Development Report 4, Vol. 1: Managing Water under Uncertainty and Risk, UNESCO, Paris, France, p. 35. https:// sustainabledevelopment.un.org/content/documents/404water.pdf (accessed 22 March 2025).

doi: 10.2166/9781789064162_0175

Chapter 6

Requested reflections

6.1 REFLECTIONS FROM AUSTRALIA

Geoffrey J. Puzon, Senior Research Scientist, Commonwealth Scientific and Industrial Research Organisation (CSIRO)
Anna Henriikka Kaksonen, Senior Principal Research Scientist, CSIRO

As the driest inhabited continent and the sixth largest country in the world, Australia has notable water management challenges. Australia's climate has warmed on average by 1.5°C since the Australian Bureau of Meteorology began collecting national records in 1910. Reduced rainfall (16–20% in southwest Australia since 1970) and increased evaporation are impacting water supply reliability, which is exacerbated by increased water demand due to population growth.

Regions of Western Australia (WA) have experienced such acute shortages of rainfall that the water flowing into dams is no longer sufficient to support local needs. Previously, Perth and regional WA relied on streamflow as their primary water source. However, the declining streamflow has resulted in the Water Corporation of WA developing a plan to diversify water sources and integrate several water supply schemes to combat the shortage. The diversified sources include desalinated seawater, groundwater, groundwater replenishment with treated wastewater and streamflow into dams, which can then supply Perth and regional WA via the Integrated Water Supply Scheme (IWSS). The IWSS allows for movement of water from various surface water catchments and sources to other parts of the drinking water distribution system (DWDS).

The rural WA DWDS covers long distances for delivering potable water to inland communities, and the water utility needs to ensure that end-point water quality remains high. The challenge in regional WA is that the DWDS consists of above-ground pipes, which are exposed to elevated temperatures that promote loss of disinfectant residual and the growth of biofilm and pathogens such as

Naegleria fowleri, Acanthamoeba and *Legionella* spp. Therefore, additional disinfection stations and monitoring may be required close to the point of use to ensure water quality for consumers. These challenges are expected to be exacerbated throughout Australia, and potentially globally, by increasing temperatures resulting from climate change.

6.2 REFLECTIONS FROM BRAZIL: FINDING NUANCES IN WELL-ESTABLISHED MYTHS

Dr. Léo Heller, Researcher, Oswaldo Cruz Foundation, Rio de Janeiro

The authors of this book engage in a necessary and courageous discussion, challenging the validity of some well-established 'truths' in the water sector. By adopting a critical perspective, they provide alternative interpretations and nuances often overlooked in a field dominated by technocratic and positivistic approaches. This is a significant contribution, as the water sector – particularly water engineering – has historically been resistant to contestation. Much of the existing literature presents fixed rules and standardized solutions that inadequately address uncertainties, multiple explanations for phenomena, and broader epistemological considerations.

Several issues raised in the book resonate deeply with my own concerns. For instance, the prevailing emphasis on centralised piped water and sewerage systems as the 'gold standard' for service provision in both urban and rural contexts is critically examined. While these systems are often promoted as ideal solutions, they can lead to significant environmental and public health externalities. Moreover, they may not always align with a human rights-based approach to water and sanitation, prioritising equality, affordability and accessibility for all.

The book also effectively critiques the mystification of technology as the primary approach to address gaps in access. The authors highlight the limitations of such an approach, supported by Kranzberg's insightful observation: 'Technology is neither good nor bad; nor is it neutral'. This critique aligns with my view that a technocratic approach, which minimises the role of public policies and management in water and sanitation provision, offers a myopic perspective on what is needed to ensure universal access to these essential services.

Another important contribution of the book is its debunking of the myth that human rights are synonymous with free services. This misconception is often exploited by certain stakeholders to discredit the human rights framework as a viable guide for service provision and public policy. By addressing this issue, the authors join a growing body of scholars and activists who advocate for human rights as a foundation for equitable and sustainable water and sanitation.

Finally, I commend the authors for challenging the notion that privatisation is the optimal model for organising service provision. This is a contentious issue, as I experienced during my tenure as the UN Special Rapporteur on the Human Rights to Water and Sanitation and when I had the opportunity to elaborate on it (Heller, Cambridge University Press, 2022). The authors ground their arguments in empirical evidence, offering a perspective on the risks associated with privatisation processes for the realisation of the human rights to water and sanitation.

By addressing these critical issues, the authors make a valuable contribution to the field, encouraging readers to question entrenched assumptions.

6.3 REFLECTIONS FROM CANADA: GOOD MYTHS, BAD MYTHS

David A McDonald, Professor PhD (Political Studies), University of Toronto

In its purest state, water is clear and simple. But it can also be murky and complicated. More than just a set of molecules, water is deeply political, cultural and economic, making it prone to myths and mythologies.

Some of these stories are wonderful, ranging from the origins of different peoples to the healing qualities of the sounds of a waterfall.

Other folklore is more problematic, including those highlighted in this book. Ranging from technological myopia to commodity fetishisation, contemporary myths about water (and sewage) can be pervasive and troubling, narrowing the scope of opportunity and often trapping water within an increasingly financialised sphere.

All is not lost, however. Celebrating water's old myths can remind us of its intangible magic, while still treating it as a concrete resource. We need a mix of starry-eyed wonder and hard-nosed realism. Water is both mysterious and rational, and it is okay to engage in both.

What we cannot afford is to fall victim to false storylines. Seeing water only through an economic lens is just as dangerous as not seeing its economic importance at all.

Separating myths from powerful interests will help promote positive storylines, providing models of governance that are respectful of water in its entirety and acknowledging the desperate reality of meeting the Sustainable Development Goals.

Hopefully this book can help to shine light on both the clear and murky pools of water in the future.

6.4 REFLECTIONS FROM CHINA

Angie Jin ZHANG, Associate Professor, Vice Director of African Studies Centre in Shanghai Normal University, China

'Dispelling Myths about Water Services' is a compelling and timely exploration of misconceptions surrounding water management. The book systematically dismantles 21 prevalent myths, offering a well-researched perspective on the challenges and complexities of water services. While rooted in Finnish data, it extends its analysis globally, making it a valuable reference for water policy scholars and practitioners worldwide.

The book's Nordic context, while insightful, invites complementary perspectives from China's rapid urbanisation. Initiatives like 'sponge cities' (using permeable infrastructure for stormwater absorption) and advanced wastewater recycling exemplify localised innovations that enrich global discourse. Yet, as the book argues, institutional reforms – such as strengthening transboundary water governance in the Yangtze River Basin or enforcing the Ecological Protection Red Line policy – remain critical to achieving the ultimate vision of 'safe, life-sustaining, and harmonious' water ecosystems.

As a researcher of water history in China, I highly appreciate *Dispelling Myths* for its clear and accessible analysis. Although the book does not fully address some of the key challenges faced by middle- or low-income countries, such as water affordability and informal water markets, it effectively dispels misconceptions and fosters a deeper understanding of critical water issues. This makes it an essential read for policymakers, scholars, and the general public.

6.5 THE CASE OF SOUTH AFRICA: A TALE OF CLIMATE CHANGE, POOR GOVERNANCE AND CORRUPTION

Prof Sadhana Manik, University of KwaZulu-Natal, Durban

Whilst it is well acknowledged that water supports life, it is equally understood that water has the potential to be the cause of destruction and disaster. These diametrically opposed sides and new caveats to the water discourse have been prevalent in South Africa where a combination of climate change, poor governance and spiralling corruption have resulted in a continued failure by government to deliver on basic services to citizens.

The World Weather Attribution service established that the likelihood of a flood event has now doubled due to climate change. Occurrences of flood events, landslides and the loss of lives and destruction to property have become frequent. Structural inequalities in society have disproportionately affected people, with the poor and marginalised in society (such as immigrants) engaging in risky behaviour such as living in informal settlements along flood plains and bearing the brunt of natural disasters.

Although the constitution of South Africa makes provision for the right to access sufficient water for domestic use, another challenge has been that of water shortages, which have been rife in both urban and rural areas where citizens have repeatedly failed to access piped water. The prevalence of a 'water and diesel mafia' have exacerbated concerns over the supply and demand for water. Where there is access to piped water, there are fears about water quality and rising costs of clean water. The supply of clean piped water is dependent on effective management of water resources, and this appears to be lacking in a majority of municipalities which are experiencing ongoing challenges.

An instrumental case of one municipality is instructional: complaints by citizens to the South African Human Rights Commission (SAHRC) about the Ethekwini Metropolitan Municipality in Durban revealed a lack of strategic planning by the municipality, insufficient water experts to provide a business plan, and a more than 50% unemployment rate in the sector. The Commission resolved that there was a failure to address a crisis in a logical manner, craft a plan of action and implement it. Other findings indicated inadequate drainage provision and the closure of weather monitoring stations due to a lack of funding. Ultimately, hazard preparedness and intersectoral planning in addition to weeding out corruption and blatant bad governance are crucial to address current water challenges in South Africa.

6.6 REFLECTIONS FROM A CAREER AT THE WORLD BANK

Mili Varughese, Senior Operations Officer, World Bank

This book offers a timely and thought-provoking examination of the myths surrounding water and sanitation service delivery. The discussion of misconceptions resonates deeply with my experiences in the sector, particularly the assumptions that water services should be free or that access is solely about infrastructure. These oversimplifications often lead to poor policy decisions, inefficient investments, and misaligned public expectations.

The book's emphasis on governance, financing, and long-term planning is especially relevant in parts of South Asia, where fragmented institutional arrangements and weak regulatory frameworks hinder service delivery. Strengthening regulatory governance—along with ensuring adequate resources—is particularly critical in countries like Ethiopia and Jordan, where maintaining continuity and resilience in water supply under conditions of scarcity remains a significant challenge.

One of the book's most important insights is the recognition that achieving SDG 6 requires both accurate cost assessments and sustainable financing strategies. Too often, the true costs of universal water access are underestimated, resulting in underfunded initiatives. The book makes a strong case for sustainable financing as a key factor in closing these gaps and ensuring long-term success.

The book also highlights how misperceptions can lead to dangerous consequences—one example being the widespread belief that groundwater is an infinite resource. In India, this misconception led to over-extraction and severe depletion of the groundwater table, underscoring the urgent need for both technological and institutional interventions to ensure long-term viability. Another invaluable insight is the principle of subsidiarity—managing water at the most locally appropriate level. From transboundary waters in the Nile basin to municipal service delivery, failures often stem from neglecting local knowledge and community perspectives.

The authors' call for expanding educational programs to include leadership, financial planning, and cross-sectoral collaboration is another critical takeaway. Too often, technical training dominates the sector, leaving gaps in areas most essential for effective water management. The book's recommendation to allocate a percentage of water utilities' turnover to research and development is noteworthy. Innovation, backed by sustained investment, is crucial to addressing the challenges of both urban and rural water service provision.

The book's holistic approach to water storage is especially valuable. Storage is often treated purely as an infrastructure issue, yet its solutions require governance, financing, and policy incentives. The book rightly positions water storage as part of a broader service delivery framework rather than a stand-alone issue. Expanding on this, storage should also be viewed through an integrated systems approach, bridging the natural, built, and hybrid continuum.

In conclusion, this book provides a comprehensive and globally relevant foundation for policymakers, practitioners, and researchers. Its insights—particularly on governance, education, financing, and integrated water management—are essential for achieving resilient and sustainable water services worldwide.

6.7 SYNOPSIS BY THE AUTHORS

As shown by the reflections each case seems to have its own challenges related to local conditions and institutional framework. At the same time there are obviously more general principles that are hopefully based more on realities rather than myths. These include at least the following issues:

- Climate, including the changes, and the need for using diverse water sources for integration of demand side management and supply side management.
- Human right-based approach as access to services rather than delivering them for free.
- Technology is to be understood in its wider context including users' sphere as well as management, institutional and governance aspects.
- Evidence of the problems of water services privatisation.
- Seeing water services not merely as an economic good but considering its other political, social, environmental and legislative dimensions.
- The importance of full costs recovery and polluter-pays principles to ensuring the sustainability of services and the resilience of services producers.

All in all, the role of institutions – rules of game in the field of water services are to be developed, updated and controlled.

doi: 10.2166/9781789064162_0181

Chapter 7
Concluding remarks

In this book, the authors examined questions that people often find problematic or intriguing regarding community water and wastewater services.

7.1 DISCUSSION

Below, we reply to the three research questions presented in chapter one.

(i) What kind of myths prevail among the public regarding water and sanitation services, and why?

True or false? The authors acknowledge that not all myths are entirely false. In the second chapter, they examine myths related to water and sanitation technologies, including: (1) that water resources and water services are synonymous, (4) that only a tap is needed, (5) that water and wastewater treatment is expensive, (7) that water towers are outdated, (9) that anything can be flushed down the toilet bowl, and (14) that water services should be free. These myths are *clearly not true*. The following myths: (10) the older the water pipes and sewers, the worse their condition, and (13) water services have too many stakeholders, hold some truth. Myth 19, The poor cannot afford water services is also likely to be true, although it has been misinterpreted through arguments like the one for free water. Most other myths, based on research findings, hardly hold any truth.

Interestingly, the myths from the authors' original Finnish study seem to be more widely applicable to the international arena than initially expected. For example, Myth 3 on water veins has references from all over the world and spans a long period of time. However, the authors caution that environmental, social and economic conditions, as well as institutional arrangements, vary case by case, and therefore, straightforward conclusions should not be drawn for specific cases. Myths may eventually be debunked, and when that happens, it would be a positive development.

(ii) How do myths impact on regulatory governance, provision and production of water services?

As a fundamental principle, citizen-oriented water services should be provided and produced for all, ensuring equal access to water and sanitation. Given the long-term implications of earlier decisions, it is essential to fully understand the available options to avoid biased decisions and policies.

To facilitate this, the water services sector should adopt the best available and most appropriate technologies, encompassing technological artefacts, systems, and know-how, including digitalisation. These technologies should prioritise energy efficiency, taking into account the needs of a circular economy. Furthermore, the actual costs involved, including renovation and renewal costs, must be covered. This necessitates long-term development and investment planning, as well as efforts in security, continuity management, and supply security.

Through multi-level governance of water services, resilience can be increased, and vulnerability reduced. The relevant public authorities will require adequate resources for regulatory governance and provision.

(iii) How could better understanding of myths be achieved in order to clarify and dispel them?

Decisions on water services policy, institutions and governance are made by politically elected persons at all levels, and particularly at the level of local governments and related bodies. We should begin systematic awareness raising programmes and campaigns to involve these parties and citizens. The aims should be to build better trust. To address this challenge, water services education at universities should be expanded to cover leadership, asset management, strategies, financial management, and best practices. Additionally, better collaboration and incentives should be created to foster inter-, multi-, and transdisciplinary understanding, ultimately meeting social needs more effectively, efficiently and cost-effectively. Incentives should also be provided for educators and researchers to publish their findings in formats that promote public awareness.

Influential water services research requires more resources than current practices allocate. To facilitate this, water utilities, as primary service producers, should be allowed and encouraged to allocate a percentage of their turnover (e.g. 1%) for research purposes. This approach should also be applied to international development collaboration.

7.2 CONCLUSIONS

Based on the authors' survey and writings, the following propositions to improve the water services in future are presented:

(i) Worldwide, the current civil and environmental engineering education and research programmes on water services have focused on water and wastewater treatment. However, these programmes should be expanded to cover management per se, and institutional, policy, and governance affairs (MIPGA) to better address the real needs of society.

(ii) Water utilities and societies must allocate more resources to Research, Development and Innovation (RDI) activities to meet the current and future challenges of sustainable and resilient water services.

(iii) Awareness raising and proactive communication by water utilities and other stakeholders should be a key strategic objective.

(iv) To support better development efforts in low- and middle-income countries, international support, technical assistance and partnerships must prioritise effective and targeted capacity-building to significantly enhance the weak absorptive capacity of the Water and Wastewater Services (WWS) sector.

Water services provision and production should be based on community water use, wastewater management, environmental protection, and overall welfare. The development of water services must consider the unique features of these services, rather than being dictated by other sectors. In addition to water quantity and quality, the priorities of water use purposes must be acknowledged, with community water supply being the foremost priority in all societies.

Water supply and wastewater services form a vital part of community infrastructure, security, and wellbeing. If water services production is compromised, the foundation of society is at risk, and people's health may be endangered. Despite these threats and challenges, the authors remain optimistic: there are many good examples and practices that need to be scaled up and utilised more widely to achieve the current SDGs and those set after 2030.

doi: 10.2166/9781789064162_0185

Key Concepts

capex	capital expenditure
continuity management	a process that helps organisations ensure that their critical functions will continue to operate in the event of an unexpected disruption
dowsing	the assumption that groundwater would occur in geological rock formations that resemble veins
dual system	separate pipelines for high quality drinking and lower quality water
economies-of-scale	advantages that can sometimes occur as a result of increasing the size of a business
green economy	aims at reducing environmental risks and ecological scarcities, and for sustainable development without degrading the environment
insdiversity	institutional diversity
institutional memories	accumulated experience and knowledge over time
institutions	the *rules of the game of a society or more formally the humanly devised constraints that structure human interaction* soccer analogy; while institutions are formal and informal rules of the game, organisations are like players
limnology	study of inland aquatic ecosystems
Litorina Sea	period of the Baltic Sea approximately 8500–4000 years ago
managed aquifer recharge	intentional recharge of water to aquifers for subsequent use or environmental benefit (earlier called artificial recharge)
myth	a long-lasting belief that is at least questionable from the point of research-based evidence

narrow corridor	how societies maintain a balance between state power and citizen rights through inclusive institutions (Acemoglu and Robinson, 2019)
natural discharge	estimated loading without the impacts of human activities
natural monopoly	where it is not feasible to construct more than one network in an area
non-point loading	loadings from dispersed sources
operative management	timeframe up to one year
outsourcing	practice of hiring a party outside a company to perform services or create goods
path dependence	the influence of past decisions on future societal processes
positivism	natural science approach largely based on quantitative measures
post-truth	circumstances in which people respond more to feelings and beliefs than to facts
pracademic research	provides scientific research answers to questions arising from society
remunicipalisation	return of previously privatised water supply and sanitation services to municipal authorities
strategic management	timeframe up to 10 years
security of supply	preparation for potential crises and disruptions and continuity management by way of safeguarding critical functions so that society, the private sector and the population can continue to operate safely
specific water use	litres per capita per day in a community
stormwater	rainwater, snow melt, and drainage water from the foundations of buildings
subsidiarity principle	service delivery should take place at the lowest appropriate governance level
tacit knowledge	knowledge, skills and abilities through experience that is often difficult to put into words or to otherwise communicate
thinking slow	to think long enough about something and thereafter have more positive and constructive views (Kahneman 2011, pp.282–283)
visionary management	timeframe up to 30 or even 50 years
waste	substance in a wrong place (Douglas, 1966)
water services privatisation	private ownership, long-term concession or operational contract
water use purpose	community water supply, nature conservation, hydropower, industrial water supply, irrigation, flood control and drainage, fishery and fish farming, recipient of wastewater, recreational use, waterborne transportation
wicked problem	a complicated problem that does not have a simple or univocal answer

doi: 10.2166/9781789064162_0187

Abbreviations

ACIR	Advisory Commission on Intergovernmental Relations
ASCE	American Society of Civil Engineers
ASR	Aquifer storage and recovery
AWWA	American Water Works Association
BOD	Biochemical Oxygen Demand (organic loading)
CADWES	Capacity Development in Water and Environmental Services
DANVA	Danish Water and Wastewater Association
EBRD	European Bank of Reconstruction and Development
EPA	Environment Protection Agency (USA)
EPSU	European Federation of Public Service Unions
EurEau	European Federation of National Associations of Water Services
FCR	full cost recovery
FIWA	Finnish Water Utilities Association
FR	futures research
GWP	Global Water Partnership
HR	historical research
HRC	Human Rights Council
HSY	Helsinki Region Environmental Services Authority
IDWSSD	International Drinking Water Supply and Sanitation Decade 1981–1990
IUWM	Integrated Urban Water Management
IWRM	Integrated Water Resource Management
LTS	Large Technological Systems
MAR	managed aquifer recharge
MDG	Millennium Development Goal
MIPGA	Management, Institutions, Policy and Governance Affairs
MSP	Municipal Services Project

NESA	National Emergency Supply Agency (Finland)
NPM	New Public Management
OECD	Organisation for Economic Co-operation and Development
OHCHR	United Nations Office of the High Commissioner for Human Rights
O&M	operations and maintenance
PFAS	per- and polyfluoroalkyl substances
PSIRU	Public Services International Research Unit
ROTI	Finnish assessment of the state of infrastructure
RWSN	Rural Water Supply Network
SCOT	Social Construction of Technology
SDGs	Sustainable Development Goals
SDG6	Clean Water and Sanitation
SHWU	specific household water use
SIWI	Stockholm International Water Institute
SWU	specific water use
THL	The Finnish Institute for Health and Welfare
TNI	Transnational Institute Multinationals Observatory
UCLG	United Cities and Local Governments
UN	United Nations
UNCED	United Nations Conference on Environment and Development (1992)
UNDESA	United Nations Department of Economic and Social Affairs
UNEP	United Nations Environment Programme
UNESCO	United Nations Educational, Scientific and Cultural Organization
UN-HABITAT	United Nations Human Settlements Programme
UNICEF	United Nations International Children's Emergency Fund
UNU	United Nations University
UNU–INWEH	United Nations University; Institute for Water, Environment and Health
UNW-DPAC	UN-Water Decade Programme on Advocacy and Communication
VAT	Value Added Tax
WASH	water, sanitation and health
WASLED	Water Services Leadership and Development
WASREB	Water Services Regulatory Board (Kenya)
WFD	Water Framework Directive (EU)
WHO	World Health Organization
WSSCC	Water Supply and Sanitation Collaborative Council
WWAP	World Water Assessment Programme
WWTP	wastewater treatment plant

doi: 10.2166/9781789064162_0189

Contributors

Authors Jarmo Hukka, Petri Juuti, Riikka Juuti and Tapio Katko are from Tampere University (TUNI), Faculty of Built Environment, CADWES research team (www.cadwes.com). Professor Eric Nealer is from the University of South Africa (UNISA) in Pretoria, RSA.

Jarmo J. Hukka is a visiting senior research fellow, an adjunct professor and a non-fiction writer. His research interests: water services regulatory governance, provision and production; privatisation; asset management; green economy; critical infrastructure protection, vulnerability and resilience. He has also worked 12 years overseas (the Cayman Islands, Sri Lanka, Kenya, Kosovo, the African Development Bank, Tunis).

Petri S. Juuti, Non-fiction writer, PhD. Research Manager and Associate Professor at three Finnish universities. UNESCO Chair in Sustainable Water Services since 2020, and the head of the CADWES research team. Guest Professor, Hubei University, China (2019–). Visiting Professor at North-West University in the 2010s and 2024–2030, University of Johannesburg (2007–13) and at the University of South Africa (2017–2021). Major interests: environmental history, water history, urban environment, city-service development, pollution, public policy and political history.

Riikka P. Juuti. DTech (2009), UNESCO Co-Chair in Sustainable Water Services and Associate Professor (Water Services, 2020–). Visiting Research Professor at the University of South Africa (UNISA, 2017–2021, and North-West University 2024–2030). Research interests: water and environmental services, sustainable water and sanitation services, water services management and environmental history. A member of the executive council of International Water History Association (2019–).

Tapio S. Katko is an Associate Professor, senior visiting expert, non-fiction author, and former UNESCO Chair in Sustainable Water Services 2012–20. His research interests are the long-term development of water infrastructure as well as institutions and governance of water services. He has worked overseas for more than five years and been involved in several international projects. He has received six international and seven national awards.

Eric J. Nealer studied and worked as a Hydro-Geological Control Technician for 17 years before he joined UNISA in 1991 until he retired as Professor in 2020. His past and current research interests are to explore gold mining, dolomite rock and sloppy stormwater and sanitation management in human settlements and municipalities in the East and West Rand, Gauteng Province of South Africa. He has successfully supervised 64 Masters and Doctoral postgraduate students.

Illustrator: **Pertti O. Väyrynen** is a graphic designer and adjustment and measurement engineer, specialising in hydraulics and pneumatics. Since 1995, he has been trained in multimedia production and graphical design and has produced technical animations. In 1999–2001, he underwent continuing education in San Francisco and New York in visualisation of animation, coding and 'writing for the web'.

Index

IWA Publishing's authorised EU representative for General Product
Safety Regulations is Diane D'Arras, 15 rue Duret, 75116 Paris,
France, e-mail: safety@iwap.co.uk.

Printed and bound by CPI Group (UK) Ltd, Croydon, CR0 4YY
05/05/2026
02102898-0006